A PROJECT MANAGER'S BOOK OF TOOLS AND TECHNIQUES

A PROJECT MANAGER'S BOOK OF TOOLS AND TECHNIQUES

A Companion to the PMBOK® Guide – Sixth Edition

Cynthia Snyder Dionisio

For general information about our other products and services, please contact our Customer Care Department within the United States at (800) 762–2974, outside the United States at (317) 572–3993 or fax (317) 572–4002.

Wiley publishes in a variety of print and electronic formats and by print-on-demand. Some material included with standard print versions of this book may not be included in e-books or in print-on-demand. If this book refers to media such as a CD or DVD that is not included in the version you purchased, you may download this material at http://booksupport.wiley.com. For more information about Wiley products, visit www.wiley.com.

Library of Congress Cataloging-in-Publication Data:

Names: Snyder, Cynthia, 1962- author.
Title: A project manager's book of tools and techniques : a companion to the
   PMBOK Guide / Cynthia Snyder.
Description: Hoboken : Wiley, 2018. | Includes index. |
Identifiers: LCCN 2017036476 (print) | LCCN 2017057182 (ebook) | ISBN
   9781119424840 (epdf) | ISBN 9781119424857 (epub) | ISBN 9781119423966
   (paperback)
Subjects: LCSH: Project management. | BISAC: TECHNOLOGY & ENGINEERING /
   Industrial Engineering.
Classification: LCC HD69.P75 (ebook) | LCC HD69.P75 S616 2018 (print) | DDC
   658.4/04—dc23
LC record available at https://lccn.loc.gov/2017036476

Printed in the United States of America

10 9 8 7 6 5 4 3 2 1

Contents

Acknowledgments

It was a wonderful experience working on this book. I have worked in project management a long time and there are still tools and techniques I needed to research. It is so great to always have new things to learn!

I was fortunate to have Richard Avery as my technical editor. His input helped make this book more approachable. His technical and interpersonal skills are among the best. Richard, I appreciate your feedback and more importantly, your friendship.

My passion for project management was only heightened by working with my team members on the *PMBOK® Guide* – Sixth Edition. My vice chair, David Hillson, helped me with many of the quantitative risk techniques in this book. I have brought my perspective to them, so for any risk gurus out there, any errors are on me, not David.

Larkland Brown and Guy Schleffer helped me with the Agile forms. They are masters at working in both the traditional and Agile worlds. Lovely Lynda Bourne is the go-to person for anything having to do with Stakeholder Engagement. I always appreciate her generous input and support.

Mercedes, Alejandro, Pan, Gwen, Mike, Kristin, and Roberta—your voices are always in the back of my mind when I write. Thank you for a wonderful experience in developing the Sixth Edition.

I so appreciate the support, friendship, and love from my husband, Dexter Dionisio. You make every day a joy.

The wonderful folks at Wiley are always a delight to work with. I feel so fortunate to have Margaret Cummins as an editor and a friend. Kalli Schultea, Lauren Freestone, Lauren Olesky, and Kerstin Nasdeo are wonderfully supportive. I am grateful for all you do.

I appreciate Donn Greenburg, Barbara Walsh, Amy Goretzky, and Roberta Storer for the work you all do to support this book and the other publications we work on together.

Thank you to all who purchase this book. I hope it brings clarity and understanding to the multiple tools and techniques we use to manage projects. May all your projects have a CPI of 1.0!

Introduction

Audience

This book is written for practicing project managers, project management students, and for those studying for the Project Management Professional certification (PMP®). The book is meant to help clarify and explain some of the more common techniques we employ in project management. It also describes some of the more specialized techniques that are not used as often, but that can be very useful in certain situations.

If you are a practicing project manager you may find it useful to read up on specific techniques to get a deeper understanding of how to apply them. You may want to find out more about a whole category of techniques, such as data representation or estimating.

If you are a student of project management you can use this book to help you understand techniques presented in class and how to apply them.

Professionals studying for the PMP will benefit by gaining an in-depth understanding of many of the techniques you will find on the exam.

What's in This Book

There are more than 125 tools and techniques mentioned in the *PMBOK® Guide* – Sixth Edition. Many of them are in one of these six groups:

- Data gathering
- Data analysis
- Data representation
- Communication skills
- Decision-making skills
- Interpersonal and team skills

This book uses some of the same categories, but not all. This book also adds a group of techniques we call Estimating. Techniques that are not in any category are put into a section called "Other." You will see the following categories in this book:

- Data gathering
- Data analysis
- Data representation
- Estimating
- Interpersonal and team skills
- Other techniques

You will not see all the 125+ techniques that are in the *Guide to the Project Management Body of Knowledge* (*PMBOK® Guide*) described in this book. Some of the techniques are just too vague to describe, such as expert judgment, quality improvement methods, or meetings. Some techniques are so descriptive that you don't really need anyone to explain them, such as ground rules, financing, or feedback. Some techniques are not included because they are general management techniques, for which volumes have already been written, for example, leadership, negotiation, and team building.

What you will find in this book is a description of 57 techniques that are used in managing projects. Some of them are used on almost every project, such as analogous estimating and rolling-wave planning. Others are more specialized, such as what-if analysis and the to-complete performance index. The techniques in this book are focused on predictive life cycles; in other words, we did not include Agile or adaptive techniques. This decision was made based on surveying potential users who rated the Agile techniques as low value, because Agile techniques are best described in a book that is dedicated to the topic of Agile.

Structure

Each section starts out with an introduction. The techniques are then presented in alphabetical order. They are not shown in the order you would build on to learn about them. For example, to learn about scheduling you would probably read in this sequence:

1. Precedence diagramming method (Section 6.6)
2. Leads and lags (Section 6.5)
3. Some, or all, of the duration estimating methods (Sections 4.1, 4.2, 4.5, and 4.7)
4. Critical path method (Section 6.2)
5. Resource optimization (Section 6.9)
6. Schedule compression (Section 6.11)

However, we needed to have a consistent way to present information, and alphabetically seemed like the best option, given the different people who will buy this book.

Each technique starts with a description of what it is. Following this are step-by-step instructions on how to use the technique. After the instructions you will see an example of how it can be used. There is some additional information, and then a listing of where you will see the technique used in the *PMBOK® Guide*.

The examples are all based around one of eight scenarios. There is an appendix in the back of the book that provides a brief overview of each scenario. The scenarios are:

- Constructing a childcare center
- Meeting company growth needs
- Developing an exam preparation video
- Developing an intranet website
- Improving the process for an IT Help Desk
- Putting in a new backyard
- Expanding a PMO information system
- Constructing a community medical center

Some of the scenarios reference previous techniques, especially those that explain earned value techniques. You don't have to read them in order, but if you are new to the technique, the previous sections are useful to help you understand the background information.

Every technique presented in this book can and should be tailored to your specific project, in your specific environment, and your specific organization. No two projects are alike. Use your experience to help you tailor the approach and the techniques to meet your needs.

For Lecture Slides of the Tools and Techniques, go to http://www.wiley.com/go/pmtools.

Good luck and may all your variances be positive!

Part 1

Data Gathering

1.0 DATA GATHERING TECHNIQUES

Data gathering is often the first technique we employ in a process. Before we can transform process inputs into outputs we often need to gather additional information. In this context, an input is any document, information, or other item that is needed to conduct a process. An output is any document, information, product, or other item that is the result of a process.

Some of the data gathering techniques entail collecting data from individuals or groups, such as focus groups and brainstorming. Some techniques entail collecting information by using tools such as checklists and check sheets. Benchmarking and statistical sampling collect data from procedures that have been performed many, many times by multiple people or even machines.

The techniques described in this section include:

- Benchmarking
- Brainstorming
- Check sheets
- Checklists
- Focus groups
- Statistical sampling

There are other data gathering techniques that are not described in this book because they are either not project specific, or they are in common use. For example, interviewing people, reviewing lessons learned and information from previous projects, and developing questionnaires and surveys are common techniques that are not applied in projects any differently than they would be in any other field. They are common enough that they don't require further explanation or examples to clarify how they are employed.

As with all techniques, you can use multiple methods to gather data. Use the methods that are easiest to gather the most complete and accurate information you need for your project.

1.1 BENCHMARKING

WHAT IT IS

Benchmarking is gathering data on the best in class, best in industry, or best in organization practices, processes, and products. The information is used as a target to improve processes, products, and results. Benchmarking is most often used in projects to collect requirements, establish quality metrics, establish cost and schedule targets, and establish stakeholder (especially customer) satisfaction targets.

HOW TO USE IT

Use the steps below as a guideline. Tailor the steps as necessary to work within your environment.

1. Identify the process or metric that you want to improve. If you are developing a new product or improving an existing one, identify the aspect of the product you are gathering data about.
2. If you are using benchmarks for improving performance, take a baseline measurement of your current performance.
3. Identify areas in your own company, leaders in the industry, or leaders in other industries with similar processes. These are your "targets" to measure against. You may need to talk with industry peers, consultants, vendors, associations, or other resources to help you discover the industry leaders (targets).
4. Depending on who and where your target is, and what you are benchmarking, you may be able to easily identify the best practices, such as when you are using a target that is internal to your organization. If the target is external, or even a competitor, you may need to gather business intelligence, work with a consultant, or find some other way to determine how they achieve their performance.

This is where the data gathering aspect of benchmarking stops. If you are working on a process improvement project that is built around achieving the benchmarks, you would develop a plan to implement a process that would help you reach the benchmark. If you are using the information for collecting requirements, the benchmark will provide information that will be prioritized along with other requirements.

Throughout the project, you can track how you are doing against the benchmark, especially if it is an easily measurable metric, such as cost per unit, time to produce, or quality defects.

Scenario: Your project is to help improve customer satisfaction with the phone support from the IT Help Desk.

You start your project by looking at the data from the satisfaction survey. You notice that the IT area with the lowest scores is the Help Desk, specifically anything to do with calling the Help Desk.

Employee Satisfaction Survey

Please rate your degree of satisfaction on a scale of 1 to 5. 1 = very dissatisfied and 5 = very satisfied.

Information Technology

Help Desk

5	4	3	2	1
How satisfied are you with the quality of service from the Help Desk?				
0.15	0.3	0.4	0.1	0.05
How satisfied were you with the hold time when you called the Help Desk?				
0.05	0.1	0.3	0.4	0.15
How satisfied were you with the ability of the person to resolve your issue?				
0.62	0.2	0.1	0.08	0
How satisfied were you with the timeliness of the issue resolution?				
0.5	0.25	0.15	0.1	0
Do you feel the person on the phone understood your issue?				
0.65	0.2	0.1	0.05	0
Did the person on the phone listen to you?				
0.45	0.25	0.15	0.1	0.05
Could you understand what the person was saying?				
0.55	0.2	0.2	0.05	0
If you left a message, was the return call timely?				
0.25	0.25	0.35	0.1	0.05

Security

Hotline

5	4	3	2	1
How satisfied are you with the quality of service from the hotline?				
0.82	0.15	0.03	0	0
How satisfied were you with the hold time when you called the hotline?				
0.75	0.15	0.1		
How satisfied were you with the ability of the person to resolve your issue?				
0.8	0.15	0.05	0	0
How satisfied were you with the timeliness of the issue resolution?				
0.87	0.1	0.03	0	0

Do you feel the person on the phone understood your issue?

| 0.8 | 0.1 | 0.1 | 0 | 0 |

Did the person on the phone listen to you?

| 0.85 | 0.05 | 0.05 | 0.05 | |

Could you understand what the person was saying?

| 0.8 | 0.1 | 0.1 | 0 | 0 |

If you left a message, was the return call timely?

| 0.9 | 0.05 | 0.05 | 0 | 0 |

The satisfaction data shows only 62 percent of the people rate their experience in calling the Help Desk as a 4 or 5 on a scale of 1 to 5. You notice the security department has the highest satisfaction: 93 percent rate their experience as a 4 or 5 when calling the security hotline. You decide to use the security hotline as the benchmark for helping to improve the IT Help Desk practices.

Additional Information

This technique is often used with focus groups (Section 1.5), market research, academic research, surveys, and questionnaires.

PMBOK® Guide – Sixth Edition References

5.2 Collect Requirements
8.1 Plan Quality Management
13.2 Plan Stakeholder Management

1.2 BRAINSTORMING

WHAT IT IS

Brainstorming is used as a technique for generating ideas and options and solving problems. It is generally a group activity that has a facilitator to manage the process. Brainstorming focuses on generating quantity, with the assumption that having a good quantity of ideas will lead to having a good-quality outcome. In the brainstorming session there is no criticism, all ideas are equal, and all are welcome, including those that seem like they are wild ideas. Various techniques can then be used to evaluate ideas gathered in order to help identify viable alternatives. Often during the process ideas will build on each other and the synergy of the group will produce better results than one person, alone, could come up with.

HOW TO USE IT

Use the steps below as a guideline. Tailor the steps as necessary to work within your environment or to work with the brainstorming variation you choose.

1. Identify the problem, goal, or outcome for the brainstorming session.
2. All members of the group state their ideas.
3. Record all ideas.

If time allows, the ideas can be elaborated, analyzed, or prioritized.

Scenario: You are managing a project to develop a new company intranet site.

As the project manager you want to get some ideas for content, design, and requirements. You bring in people from various departments to help brainstorm some ideas that you will later prioritize and send out for the rest of the organization to comment on.

You decide to use some variations on the traditional brainstorming technique by using a group passing technique, some electronic brainstorming, and individual brainstorming.

Group passing technique. To gather high-level requirements, you give each person a category of requirements to work on. You ask each person to write down his or her requirements and then pass it to the next person, who adds his or her requirements. This continues until everyone has commented on each category.

Individual brainstorming. Rather than being done in a group, individual brainstorming is done as an individual. This can take the form of free-writing, free-speaking, or drawing a mind map (Section 3.6). You ask your brainstorming group to come up with some ideas for the design and graphical user interface (GUI) for the website. You tell them they can use free-writing, picture creation, mind mapping, or any other form of communication for their ideas.

Electronic brainstorming. In this brainstorming technique you ask the group to contribute in an online environment. You post the content topic in an online bulletin board or chat room and people respond. Electronic brainstorming can permit a large number of ideas to be gathered very quickly because there is no turn-taking. People can respond as soon as they see other ideas, and the energy can build on itself. Electronic brainstorming has been found to generate more ideas and be of greater quality than in-person brainstorming. However, it requires a moderator to ensure anonymity does not lead to disrespectful interactions. Electronic brainstorming can also take place over a longer period of time, allowing for more reflection. You choose how long to allow the bulletin board to be posted, allowing people to log in and contribute as they are able.

Additional Information

Brainstorming can be used with focus groups (Section 1.5). Combining brainstorming with the nominal group technique (Section 5.3) allows the ideas to be prioritized for further elaboration or to reach a decision.

PMBOK® Guide – Sixth Edition References

4.1 Develop Project Charter
4.2 Develop Project Management Plan
5.2 Collect Requirements
8.1 Plan Quality Management
11.2 Identify Risks
13.1 Identify Stakeholders

1.3 CHECK SHEETS

WHAT IT IS

A check sheet is a tally sheet that is used to collect data. It can be used to collect data about defects or to keep track of completing steps in a process.

HOW TO USE IT

Use the steps below as a guideline. Tailor the steps as necessary to work within your environment.

1. Identify the types of defects (or other variables) you are looking to tally and enter these in Column 1 of a spreadsheet. You would start this in cell A2 and continue with A3, A4, and so on.
2. The top row can Indicate either frequency (enter "Frequency" in cell B1) or, if you are tracking locations or days of the week, hours of the day, or some other variable, enter each of these in the top row starting with cell B2. Follow with B3, B4, and so on.
3. Observe the process, outputs, or deliverables.
4. Indicate the source of the defect and put a mark in the appropriate cell.
5. Tally the rows and columns.

Scenario: Your project is to help improve customer satisfaction with the phone support from the IT Help Desk.

To understand the reason behind the IT Help Desk complaints you create a check sheet to tally the number of complaints, by reason and by department. The number of complaints by reason are totaled, as are the number of complaints by department.

	IT Help Desk	Security Hotline	Legal Info Line	Maintenance Trouble Line	Total
Call was dropped	IIIIIII	I	III	0	11
Tech did not listen to me	IIII	0	IIIII	II	11

	IT Help Desk	Security Hotline	Legal Info Line	Maintenance Trouble Line	Total
On hold too long	IIIIIIIIIIIIIIIIIIIIIII	III	IIIIII	IIIIIIII	39
Couldn't understand technician	IIIIII	II	IIII	III	15
Could not help with my issue	IIIIIIIIII	III	IIIIIIII	III	24
Did not return my call in a timely manner	IIIIIIIIIIIII	0	II	IIIII	20
Total	62	9	28	21	120

Additional Information

Check sheets can be used to show the distribution of defects, and can then be arranged in a histogram (Section 3.4) or Pareto chart showing the frequency of defects by cause, location, or other variable.

PMBOK® Guide – Sixth Edition References

8.3 Control Quality

1.4 CHECKLISTS

WHAT IT IS

A checklist is a list of activities, steps, or procedures that need to be done. It is often used as a reminder.

HOW TO USE IT

In projects, checklists are used to reduce or eliminate risks or defects. They contain a series of steps that must be taken, or processes that must be completed.

Scenario: You are managing a project to develop a new company intranet site.

This example shows some of the items that must be completed in updating an intranet website. This is only an example; all checklists should be tailored to your specific environment.

- ☑ Collect stakeholder requirements
- ☑ Map inbound links
- ☑ Create new content as necessary
- ☑ Upload content to new site
- ☑ Develop tags and metadata
- ☑ Check all links
- ☑ Create XML/HTML sitemaps
- ☑ Test speed
- ☑ Check for mobile access
- ☑ User acceptance testing

Additional Information

A few of the downfalls associated with using checklists are:

- People can rely on the checklist and fail to look outside the items on the checklist for other risks or causes of failure. Use the checklist as a starting place, not an ending place.
- Over-reliance on checklists can replace common sense or critical thinking. A checklist can act as a prompt, but should not take the place of looking at a situation and taking appropriate actions.

Therefore, checklists should only be used as a starting point in many project situations. Team members should continue to identify sources of risks and defects.

PMBOK® Guide – Sixth Edition References

4.2 Develop Project Management Plan
8.2 Manage Quality
8.3 Control Quality
11.2 Identify Risks

1.5 FOCUS GROUPS

WHAT IT IS

A focus group is a group of prequalified people who are brought together to provide information about a product, service, or result. A professional moderator uses a question guide to focus the direction of the questions. The moderator also observes behaviors and nonverbal cues and records them in her observations.

HOW TO USE IT

Focus groups are most commonly used for new product or new service development. They may be part of a market research campaign to gather requirements or to provide insight into customer opinions, expectations, desired benefits, underlying assumptions, common views, and so forth.

1. Establish your goals or desired outcomes for the focus group.
2. Develop a discussion guide that provides a focus for the group, but also allows some open conversations that are not driven by the moderator.
3. Find a qualified moderator. You may want to hire a professional moderator to lead the group. Your moderator should have at least these qualifications:
   a. The ability to be friendly, nonjudgmental, flexible, and open. He or she needs to stay on topic, ask open-ended questions, and manage the conversation so it is productive and not combative.
   b. He or she should have some understanding of the new product or service, but not have a vested interest in the outcomes.
4. Determine how to record the session, either with video, audio, or a note-taker.
5. Set up the meeting logistics such as time, location, date, and duration.
6. Identify and invite the participants. Ideally you want a cross section of potential end users and customers.
7. At the opening of the meeting:
   a. Thank participants for attending.
   b. Review the purpose of the meeting.
   c. Review the flow of the meeting, guidelines, ground rules for participation, and so forth.

8. During the meeting:
   a. Ask open-ended questions to start a topic; use closed-ended questions to gain clarity.
   b. Make sure everyone is participating. You may need to call on some people to get their feedback.
   c. Summarize information, rephrase questions if necessary, employ active listening, and ask for comments about responses.
   d. Ask if there are any other comments or thoughts before moving on, and before wrapping up the meeting.
   e. Thank the participants for coming.
9. Assess the data and look for patterns, themes, unexpected outcomes, and new questions that may have arisen.

Scenario: You are the project manager for a project to implement a childcare facility for your organization's employees.

The Project Charter has been developed. You are meeting with 12 parents at your organization who currently use childcare facilities. They work in different departments, have children from three months old to four years old, and hold a variety of jobs.

You want to use the focus group to understand the expectations and attitudes about the curriculum, play time activities, and food. You decide to be the note-taker and you have asked Roberta, the Director of Human Resources, to facilitate the session. Roberta doesn't have a personal relationship with any of the participants and she is skilled in listening, drawing people out, and creating a nonjudgmental environment.

You set up the meeting for lunch time two weeks in the future. You will provide pizza and soda for the participants. Roberta opens the meeting with general questions about what people are looking for in a childcare facility. From there she asks questions about the ratio of skill development versus playtime, the types of indoor and outdoor play time equipment the parents would like to see, and their thoughts on snacks and lunches.

As the meeting closes the parents are excited about the new childcare center and make themselves available for any future questions. The information you gained will help you put together an RFP for playground equipment and food vendors. You will provide the information on the curriculum and skill development to the director of the childcare center to help him develop the roadmap for learning.

Additional Information

Some focus groups offer incentives, including cash for participation.

Focus groups generally provide broad qualitative information. The interactions and body language provide a deeper understanding of people's opinions than questionnaires. The qualitative information can then be analyzed and can be followed up with questionnaires and surveys to get quantitative data by asking ranking questions or closed-ended questions.

PMBOK® Guide – Sixth Edition References

4.1 Develop Project Charter
4.2 Develop Project Management Plan
5.2 Collect Requirements

1.6 STATISTICAL SAMPLING

WHAT IT IS

Statistical sampling is selecting a subset of a population to estimate characteristics and information about the whole population. Selecting a subset is a more time- and cost-efficient way to reach conclusions about a group instead of asking each member of the group, or testing every physical component.

HOW TO USE IT

Statistical sampling can be used to infer behavior of a group of people, or to infer the quality of a set of deliverables or components.

1. Identify the population of interest.
2. Identify the variables you want to measure.
3. Select the sample size and method for selecting the sample population.
4. Implement the sampling plan.
5. Measure the variables from the sample.

The most effective ways to select your sample are to choose a random sample, or a systematic sample (such as every 15th person).

Scenario: You have been asked to meet the physical growth needs of Top Dog Project Services.

One of the options you are reviewing to meet the growth needs of all the offices is to develop a work-from-home program. Before establishing the requirements and policies for the program you want to get an idea of how many people work from home after hours or on weekends. You also want to see, on average, how many hours people are active on their computer during the workday. You think it would be useful to have this information broken out by department.

First you identify the locations that have employees that will be eligible to work from home. You identify four locations that have employees who would be eligible for the program. You determine that the total number of employees who would be eligible to work from home is

250 people. That is your population of interest. You decide that you need at least three people from each department as a minimum. For departments with more than 30 people, you decide to identify one person for every ten people in the department.

You meet with the Corporate Director of Information Services and tell him your sampling parameters. He agrees that these are reasonable measures and confirms that he can provide the information anonymously so that no one's personal work habits are identifiable. He runs the sample for four weeks to get an idea of activity over time.

Additional Information

When you are selecting your sample, try to ensure you don't have over-coverage or under-coverage of a particular trait or bias in your sample.

There is a wide body of information on selecting sample sizes for statistical sampling. For the most part, in project management we don't have to be extremely precise. We mostly use the information to give us a general idea of a population's characteristics. For projects that require Six Sigma–type measurements, you will need to be a lot more precise in selecting the sample and defining the variables you want to measure.

PMBOK® Guide – Sixth Edition References

8.3 Control Quality

Part 2

Data Analysis

2.0 DATA ANALYSIS TECHNIQUES

Data analysis techniques are used to assess and evaluate data in order to discover or gain deeper information about a topic. They can also be used to support decision-making. We use many different methods of data analysis in project management. Some of them are very broad, such as alternatives analysis and decision trees. Others are specific to a particular knowledge area, such as a stakeholder analysis or a make-or-buy analysis.

The techniques described in this section include:

- Alternatives analysis
- Cost benefit analysis
- Cost of quality
- Decision tree analysis
- Earned value analysis
- Influence diagrams
- Make-or-buy analysis
- Performance indexes
- Regression analysis
- Reserve analysis
- Root cause analysis
- Sensitivity analysis
- Stakeholder analysis
- SWOT analysis
- Technical performance analysis
- Variance analysis
- What-if analysis

The techniques in this group can be some of the most challenging to master. Some of them can be quite complex, such as a sensitivity analysis or earned value analysis. Understanding what project data is telling you is critical to making good decisions. Therefore, the effort you put into data analysis can have a significant impact on the success of your project.

This book does not describe data analysis techniques that are very general and easy to understand, such as document analysis or assumption and constraint analysis. It also does not cover techniques that require specialized knowledge or software, such as simulations (Monte Carlo analysis). The type of project you are working on, the availability of quality data, and in some cases specific software, will influence the data analysis techniques you can use on your project.

2.1 ALTERNATIVES ANALYSIS

WHAT IT IS

Alternatives analysis is a technique used to evaluate and select options or approaches to execute and perform project work. An alternatives analysis can be used at a high level; or you can apply a more rigorous approach that incorporates a matrix that identifies rating criteria and weighting factors. This technique can be used in several processes and in a wide range of industries.

HOW TO USE IT

The steps below describe how to conduct an alternatives analysis using a matrix to weight evaluation criteria. Use the steps below as a guideline. Tailor the steps as necessary to work within your environment. For example, you can apply this technique without weighting the evaluation criteria, in which case you would omit Step 3.2.

1. Identify the problem or decision you are evaluating. For more information about decision-making, refer to Section 5.2. For more information about problem solving, refer to Section 5.4.
2. Define Solution Requirements. Often the Solution Requirements can be extracted from Requirements documentation, the Statement of Work, the Scope Statement, or other project documents. For information technology (IT) projects, "Use Cases" can also be used to identify the Solution Requirements.
3. Define Evaluation Criteria and Weightings. The relevant stakeholders should provide input on the criteria they need in a solution or outcome. They should also provide weight values for each criterion. The project manager or business analyst often facilitates this process. The following steps are used to define and weight the selection criteria.
   3.1. Identify the selection criteria.
   3.2. Develop a "weighted value" for each criterion. It is a best practice for all weights to add up to 100 percent, but this is not mandatory.
   3.3. Define the scoring algorithms to determine how effectively each alternative meets the selection criteria. For example, if low cost of ownership is a selection criterion, define how you would rate alternatives on this criterion on a scale of 1 to 5, with 1 being the highest cost option and 5 being the lowest cost option.
   3.4. Create a Scoring Matrix with a space for alternatives on one axis and the criteria and weighting on the other axis.

4. Identify Options and Conduct Market Research. To conduct market research, you can send out a Request for Information (RFI), hold meetings with vendors, perform Internet research, or talk with consultants.
5. Conduct Initial Assessment, Score and Evaluate, and Eliminate Options. The initial assessment can be used to narrow the field to the final one or two options, or it may provide you with the best solution, in which case you skip the next step.
6. Conduct a Cost Benefit Analysis with Risk Adjusted Costs to Eliminate More Options. This step is used to evaluate the highest scoring options considering the various risks that each option contains. Depending on the size of the investment, it may be appropriate to conduct an extensive risk analysis on the potential solutions before finalizing a recommendation.
7. Recommend Solution. The solution that scores the highest using the selection criteria and weighting, and has an acceptable risk profile, is presented to the stakeholder who has the authority to make the final decision. Sometimes this is the project manager, sometimes it is the sponsor or customer. The backup documentation and a summary of the process should be included with the recommendation.

The following examples demonstrate a relatively simple use of alternatives analysis, followed by a more rigorous example that incorporates weighted criteria to analyze the options.

Evaluating Time, Cost, and Resource Constraints

Scenario: You are managing a project to develop a new company intranet site.
You are analyzing whether to use in-house resources or outsource for higher-skilled resources that cost more per hour, and can accomplish the work faster. The relative importance of the schedule constraints, cost constraints, and resource availability are the solution requirements.

Option 1 assumes you have a team member with adequate skills who can accomplish a task in 80 hours. Her hourly rate is $45.

Option 2 assumes you can outsource the work and use a highly skilled resource that will only need 65 hours to accomplish the work. The higher-skilled resource has an hourly rate of $60.

By multiplying the rate times the duration you determine the cost of the in-house resource is $3,600. The cost of the outsourced, higher-skilled resource is $3,900.

If your solution requirements state that low cost is a higher priority than the schedule, or if you prefer to keep the work in-house, you should go with Option 1. If time is of the essence and you have room in the budget, you should go with Option 2.

Selecting a Cafeteria Management Service

Scenario: You are the project manager for a project to implement a childcare facility for your organization's employees.
This example focuses on selecting a vendor to provide the food service for the Childcare Center cafeteria. Weighted solution (evaluation) criteria are used to evaluate vendors against a set of requirements for the cafeteria services.

1. **Define Solution Requirements.** In this example, the requirements are:
* The service provider should have food options, with a preference toward fresh fruits and vegetables that are locally sourced.
* A reliable vendor, as determined by references.
* Pricing that is consistent with the market. Should be able to provide a healthy lunch for less than $5.
2. **Define Evaluation Criteria and Weightings.**

   The selection criteria and weights are as follows:

* Healthy food options: 50 percent
* Locally sourced where possible: 15 percent
* Good customer feedback: 25 percent
* Competitive pricing: 10 percent

   The scoring algorithms are:

Healthy Food	5 = 75% of the menu includes fruits, vegetables, grains, legumes, and lean meat. 25% or less is frozen or processed.
	4 = 66–74% of the menu includes fruits, vegetables, grains, legumes, and lean meat. 33% or less is frozen or processed.
	3 = 50–65% of the menu includes fruits, vegetables, grains, legumes, and lean meat. 50% is frozen or processed.
	2 = Most of the menu incorporates fried, frozen, or processed foods.
	1 = Limited fresh options. Almost all food is fried, frozen, or processed.
Locally Sourced	5 = More than 75% is locally sourced within 100 miles.
	4 = 66–74% is locally sourced within 100 miles.
	3 = 50–65% is locally sourced within 100 miles.
	2 = Most of the food is transported from greater than 100 miles.
	1 = Almost all the food is transported from greater than 100 miles.
Good Customer Feedback	5 = 5 out of 5 references report they are happy with the vendor and would contract with them again.
	4 = 4 out of 5 references report they are happy with the vendor and would contract with them again.
	3 = 3 out of 5 references report they are happy with the vendor and would contract with them again.
	2 = 2 out of 5 references report they are happy with the vendor and would contract with them again.
	1 = 1 out of 5 references reports he is happy with the vendor and would contract with them again.
Price	Lowest average price
	Within 5% of lowest average price
	Within 10% of lowest average price
	Within 15% of lowest average price
	Greater than 15% higher than lowest average price

(continued)

(continued)

Create a Scoring Matrix

Vendor Criterion	Weight	Rating	Score	Rating	Score	Rating	Score
Healthy	.50						
Locally Sourced	.15						
Customer Feedback	.25						
Price	.10						
Total	**1.0**						

3. **Identify Options and Conduct Market Research.** For this project you survey local businesses that provide onsite cafeterias. You find some national brands and some local vendors.
4. **Conduct Initial Assessment, Score and Evaluate, and Eliminate Options.** There are three vendors that respond to your request for proposal. You evaluate the proposals and score them based on the criteria and scoring algorithm.

Vendor Criterion	Weight	National Foods Corp		Jodi's Kitchen		Organic Options	
		Rating	Score	Rating	Score	Rating	Score
Healthy	.50	2	1	3	1.5	4	2
Locally Sourced	.15	3	.45	5	.75	4	.60
Customer Feedback	.25	4	1	3	.75	3	.75
Price	.10	5	.5	4	.40	3	.30
Total	**1.0**		**2.95**		**3.40**		**3.65**

5. **Conduct Cost Benefit Analysis with Risk Adjusted Costs and Eliminate Options.** Assume for this example that Organic Options had the highest score. You have learned that they have recently undergone a change in management. The previous management was not very responsive to customer requests. It appears that in the last four months with the new management, there has been a much higher satisfaction rating. This risk is evident by the fact that they only scored 3 for the customer feedback criteria. You decide that you can address this risk by developing a Service Level Agreement as part of the contract and that you can build in an exit strategy after six months if you are not satisfied with the customer service.
6. **Recommend Solution.** In this example, the scoring sheets and the recommendation for including the Service Level Agreement and the exit clause in the contract would be given to the decision maker.

Additional Information

Some organizations refer to the high-level identification of options as an alternatives analysis and the scoring and ranking as multi-criteria decision analysis. Alternatives analysis can be used at the start of the project to determine the best approach, during the project to assist in selecting a vendor, in quality management to identify options to respond to quality issues, and in risk management to determine an acceptable risk response.

PMBOK® Guide – Sixth Edition References

4.5 Monitor and Control Project Work

4.6 Perform Integrated Change Control

5.1 Plan Scope Management

5.3 Define Scope

6.1 Plan Schedule Management

6.4 Estimate Durations

7.1 Plan Cost Management

7.2 Estimate Costs

8.2 Manage Quality

9.2 Estimate Resources

9.6 Control Resources

11.4 Perform Quantitative Risk Analysis

13.4 Monitor Stakeholder Engagement

2.2 COST BENEFIT ANALYSIS

WHAT IT IS

A cost benefit analysis (CBA) is used to assess options that provide the best approach to achieving benefits while minimizing costs. It can be used to assess the viability of a project and to rank various approaches or alternatives for meeting project objectives.

When looking from a short-range financial perspective, actual costs and benefits are assessed. When looking from the long-term perspective, the time value of money is taken into consideration. This is accomplished by converting future payouts and expenditures into the present value. Some projects take into consideration the life cycle cost of the product; others focus on the project costs.

When assessing whether to do a project, such as when preparing a business case, the present value costs are subtracted from the present value benefits. The sum of benefits less the costs is called the net present value (NPV). If the NPV is positive, the project is expected to return a profit. The higher the NPV, the more profitable the project.

When assessing various approaches to a project, or responses to a risk, you can rank the preference of options by putting those with the highest NPV first.

HOW TO USE IT

Use the steps below as a guideline. Tailor the steps as necessary to work within your environment.

1. List the options you are evaluating.
2. Document the costs for each year, for each option.
3. Document the benefits (translated into currency) for each year, for each option.
4. Apply a discount rate (the expected cost of money) for each year.
5. Calculate the net present value by subtracting the costs from the benefits.

Scenario: Your organization is updating its current PMO information system infrastructure with all new software, cloud computing, collaboration sites, and real-time reporting software.

The Director of Marketing wants to compare the benefits of the PMO information system with developing a new line of business by creating an online project management training curriculum.

For the PMO information system the initial investment is $750,000 and $45,000 per year to maintain. The benefits would not be available until the second year. The enhanced infrastructure is expected to generate revenue, as shown in the table, for Years 2 through 5 before needing to be replaced.

The online project management training investment is $325,000 in Year 1 and $35,000 in Years 2 through 5 to maintain. You expect to have $1,500,000 in sales the first year and $3,500,000 in Years 2 through 5.

The table below shows the information presented above.

Year	PMO Upgrade			Virtual Training		
	Benefits	Costs	NPV	Benefits	Costs	NPV
1	—	750,000		1,500,000	325,000	
2	2,500,000	45,000		3,500,000	35,000	
3	5,250,000	45,000		3,500,000	35,000	
4	6,000,000	45,000		3,500,000	35,000	
5	6,000,000	45,000		3,500,000	35,000	
		Total			Total	

The discount rate is 5 percent.

The table below shows the net present value, for Years 1 and 2, with the discount of 5 percent applied starting in Year 2.

Year	PMO Upgrade			Virtual Training		
	Benefits	Costs	NPV	Benefits	Costs	NPV
1	—	750,000	(750,000)	1,500,000	325,000	1,175,000
2	2,500,000	45,000	2,338,095	3,500,000	35,000	3,300,000
3	5,250,000	45,000		3,500,000	35,000	
4	6,000,000	45,000		3,500,000	35,000	
5	6,000,000	45,000		3,500,000	35,000	
		Total			Total	

Year 1 does not apply the discount rate, since all rates are based on current year information. Thus, the NPV for Year 1 is the benefits minus the costs. Year 2 calculates the benefits minus the costs and then divides that total by 1.05 to account for the 5 percent discount rate.

Year 3 will take the benefits minus the costs and divide by 1.05^2. The following year, the divisor will be 1.05^3. The final year, the divisor will be 1.05^4. The completed table is shown below.

(continued)

(continued)

Year	PMO Upgrade Benefits	Costs	NPV	Virtual Training Benefits	Costs	NPV
1	—	750,000	(750,000)	1,500,000	325,000	1,175,000
2	2,500,000	45,000	2,338,095	3,500,000	35,000	3,300,000
3	5,250,000	45,000	4,721,088	3,500,000	35,000	3,142,857
4	6,000,000	45,000	5,144,153	3,500,000	35,000	2,993,197
5	6,000,000	45,000	4,899,193	3,500,000	35,000	2,850,664
		Total	16,352,530		Total	13,461,718

Given this information, the investment in the PMO upgrade is a better option because it has a higher net present value.

Additional Information

This technique is also known as a benefit cost analysis. You may see it used with a cost of quality analysis (Section 2.3) or a make-or-buy analysis (Section 2.7). Sometimes a sensitivity analysis (2.12) is conducted on the option selected to see if it is stable, or subject to changes in variables.

Some of the pitfalls to be aware of when conducting a cost benefit analysis include:

- Using estimates from outdated projects, or projects that are not similar, will lead to faulty estimates.
- Subjectivity or bias in cost or benefit estimates can make the estimates unreliable.
- Confirmation bias (looking for reasons to go forward with a project) can lead to more favorable estimates than are realistic.
- Exclusion of significant upfront or maintenance costs can misrepresent the outcomes.

Be aware that the cost benefit analysis only looks at the financial impact of options. It does not take into account the intangible benefits and costs such as goodwill, morale, reduced turnover, social benefits, and so forth. Thus, the CBA is one technique to consider, but it is not the only method to assess the viability of a project or an approach.

PMBOK® Guide – Sixth Edition References

4.5 Monitor and Control Project Work
4.6 Perform Integrated Change Control
8.1 Plan Quality Management
9.6 Control Resources
11.5 Plan Risk Responses

2.3 COST OF QUALITY

WHAT IT IS

Cost of quality analysis is used to find the appropriate balance for investing in quality prevention and appraisal to avoid defects or product failures. There are four categories of costs associated with quality:

- **Prevention.** Costs incurred to keep defects and failures out of a product.
- **Appraisal.** Costs incurred to determine the degree of conformance to quality requirements.
- **Internal Failure.** Costs associated with finding defects before the customer receives the product.
- **External Failure.** Costs associated with defects found after the customer has the product.

The information below summarizes the types of costs and provides examples. This information is paraphrased from the American Society for Quality (ASQ).[1]

Prevention	Prevention costs are incurred to prevent or avoid quality problems. These costs are associated with the design, implementation, and maintenance of the quality management system. They are planned and incurred before actual operation, and they could include: • Product or service requirements—establishment of specifications for incoming materials, processes, finished products, and services • Quality planning—creation of plans for quality, reliability, operations, production, and inspection • Quality assurance—creation and maintenance of the quality system • Training—development, preparation, and maintenance of programs
Appraisal	Appraisal costs are associated with measuring and monitoring activities related to quality. These costs are associated with the suppliers' and customers' evaluation of purchased materials, processes, products, and services to ensure that they conform to specifications. They could include: • Verification—checking of incoming material, process setup, and products against agreed specifications • Quality audits—confirmation that the quality system is functioning correctly • Supplier rating—assessment and approval of suppliers of products and services

(continued)

---

[1] http://asq.org/learn-about-quality/cost-of-quality/overview/overview.html as of March 2, 2017.

(continued)

Internal Failure	Internal failure costs are incurred to remedy defects that are discovered before the product or service is delivered to the customer. These costs are incurred when the results of work fail to reach design quality standards and are detected before they are transferred to the customer. They could include: • Waste—performance of unnecessary work or holding enough stock to account for errors, poor organization, or communication • Scrap—defective product or material that cannot be repaired, used, or sold • Rework or rectification—correction of defective material or errors • Failure analysis—activity required to establish the causes of internal product or service failure
External Failure	External failure costs are incurred to remedy defects discovered by customers. These costs occur when products or services that fail to reach design quality standards are not detected until after transfer to the customer. They could include: • Repairs and servicing—of both returned products and those in the field • Warranty claims—failed products that are replaced or services that are re-performed under a guarantee • Complaints—all work and costs associated with handling and servicing customers' complaints • Returns—handling and investigation of rejected or recalled products, including transport costs

HOW TO USE IT

The costs associated with quality are often built into an organization's infrastructure, policies, and procedures. It can be difficult to obtain that information for project analysis. Thus, for projects we usually use a cost of quality analysis when we are evaluating whether to invest more upfront for a more resilient product, or whether we want to get the work done quickly and absorb the time and cost associated with fixing issues later. This may be as simple as building in redundancy, going through an extra design or technical review, or outsourcing to a vendor to provide maintenance. Regardless of the outcome of your analysis, your decision needs to be aligned with organizational policies and stakeholder requirements.

When you are conducting a cost of quality analysis for your project, don't include the costs already associated with the quality policy of the organization. For example, if your organization hires a vendor to come out and calibrate and maintain certain pieces of equipment on an annual basis, don't include those costs in your "appraisal" costs for the project. Only include costs that are specific to the project.

1. Identify all the costs associated with preventing or avoiding defects or quality problems. This includes prevention and appraisal costs. Consider such costs as higher-quality materials, a more robust design and review process, documentation, training, inspections, and audits.
2. Identify all the costs associated with defects in the product. Include costs associated with scrap, rework, root cause analysis, warranty work, repairs, returns, and even possible lawsuits and loss of customers.
3. Sum the costs of conformance (money spent to avoid failure) and the costs of nonconformance (money spent during and after the project because of failure).
4. Choose the solution that shows the optimal balance for investing in cost of prevention and appraisal to avoid failure costs.

Scenario: Develop an eight-hour in-house training video to prepare employees for an industry certification.

The organization expects 150 employees to take the certification over the next four years. You plan on using employees who have already passed the exam as subject matter experts (SMEs) to develop the materials. You will also use your employees to present the material on video.

You assume this can be done in a classroom with two video cameras using the SMEs to present the material. You performed a 15-minute trial taping as a prototype (Section 6.8). The result was a video that had very poor sound quality. The material was acceptable and the employees did okay on camera, though they did require a few retakes to get acceptable results.

Based on the 15-minute trial taping results, you solicited several quotes from videographers, video studios, talent, editors, and directors. You will be presenting a cost summary and recommendations to the sponsor and want to justify your recommendations from a cost of quality perspective. The information in the table reflects the best value quotes based on the responses.

Cost of Conformance		Cost of Nonconformance	
Prevention		**Internal Failure**	
Video package[1] Includes director, editor, cameraman, and studio time	$12,000	Addendum[2]	$1,500
Talent	$2,400		
Appraisal		**External Failure**	
Technical Reviewer	$1,000	Re-testing[3]	$15,000

1. Assumes 16 hours of time to record 8 hours of video.
2. Assumes you would need to publish a paper addendum correcting the misinformation and distribute it to all potential exam takers.
3. Assumes a 20 percent failure rate if content or delivery is not clear. Each retake of the exam is $500.

Based on these costs you are planning on recommending the video package and the technical reviewer as an optimal investment in quality for this project. You believe that with the director and editor that, your SMEs will do a good job of presenting the information; therefore you are not recommending hiring outside talent. While you feel confident that your SMEs can develop good material, you think the investment in having an outside expert review the content is a wise decision.

You assume that with the high-quality video production you will not have any external failure costs associated with poor quality. In other words, any failures will be based on the students' lack of studying, not the company producing a poor product. This will result in the employee paying for the cost of retaking the exam. Thus, the cost of conformance is $13,000 (video package + technical reviewer) and the cost of nonconformance is $16,500.

Additional Information

A cost of quality (COQ) analysis is often used in parallel with or instead of a cost benefit analysis (Section 2.2). It may also be part of a make-or-buy analysis (Section 2.7). If you are using the COQ to evaluate multiple options, you may want to model the outcomes with a decision tree (Section 2.4).

PMBOK® Guide – Sixth Edition References

  7.2 Estimate Costs
  8.1 Plan Quality Management

2.4 DECISION TREE

WHAT IT IS

Decision tree analysis evaluates uncertainty using a diagram with branches that model various options or outcomes and their implications. In project management, we use it to evaluate which projects to invest in and which approaches to take, and to model uncertainty associated with cost, schedule, or risk events.

HOW TO USE IT

1. Identify the various scenarios or outcomes that could occur. In some situations, none of the outcomes may occur; however, the point is to develop a model that represents what could happen given the information you have at the time.
2. Create a diagram where each option is a branch.
3. Determine the probability of each scenario in each branch so that the total for each branch is 100 percent.
4. Determine the monetary value associated with each outcome.
5. Multiply the probability times the monetary value of each outcome.
6. Sum the outcomes for each branch to derive the expected monetary value of each option.

Remember, the decision tree is just a model of what *could* happen. It is not factual. It is a technique used to model uncertainty. As with any model, the fidelity is only as good as the data behind it. In other words: garbage in, garbage out.

Scenario: Your organization is updating its current PMO information system infrastructure with all new software, cloud computing, collaboration sites, and real-time reporting software.

You are evaluating whether to purchase new reporting software and customize it or to develop it in-house. This is your initial branch.

(*continued*)

(continued)

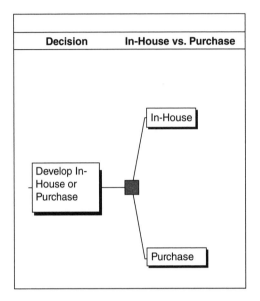

If you develop in-house, you risk losing resources to other projects. Based on the current environment, you assume there is a 50/50 chance that you'll lose 40 percent of your staff.

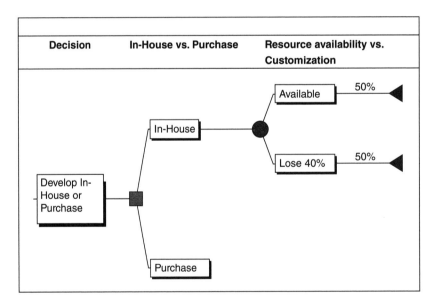

If you keep your staff, developing software will cost $172,000. If you lose staff, you have to hire outside contractors, and developing software will cost $208,000.

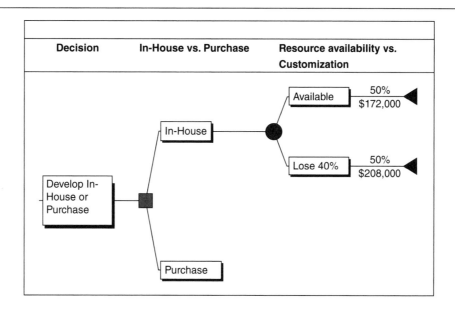

By multiplying the probability of each branch times the value of each branch you get the expected monetary value of each branch. Add the value of each branch to get the expected monetary value of developing in-house.

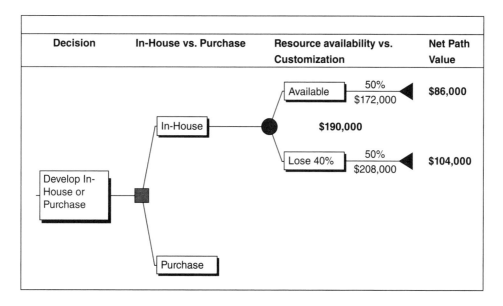

If you purchase ready-made software, there is uncertainty about the amount of customization you will have to do to get the software to work with your systems. There is a 60 percent chance that you will have to do only a little customization, which would bring the total cost to $152,000. There is a 40 percent chance that you will have to do a lot of customization, which would bring the total cost to $188,000.

(continued)

(continued)

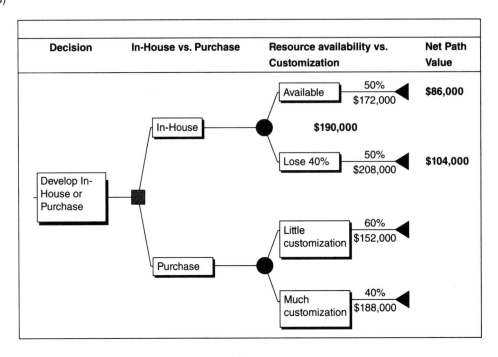

To calculate the expected value, add the values of each branch under the "purchase" option.

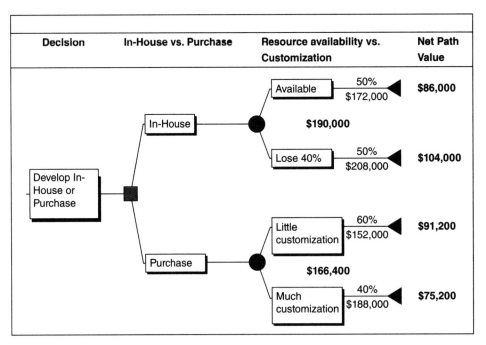

Your decision would be to purchase software because the expected monetary value is lower.

Additional Information

Remember, sometimes the monetary values and the probability estimates are not precise. To build a decision tree you need to develop a deterministic estimate, even though you may have a range of possible outcomes. A deterministic estimate is a single point estimate, rather than a range of estimates.

The decision tree used in the example has a lot of uncertainty built into it. Some decision trees have more definitive information, such as interest rates associated with purchasing versus leasing or hiring versus using contract labor.

A decision tree is often used to model "expected monetary value" information. You can also use it to model cost of quality (Section 2.3), make-or-buy decisions (Section 2.7), risk responses, and what-if scenarios (Section 2.17).

PMBOK® Guide – Sixth Edition References

11.4 Perform Quantitative Risk Analysis

2.5 EARNED VALUE ANALYSIS

WHAT IT IS

Earned value analysis is a technique that integrates scope, schedule, and cost information into single measures to assess the status of project performance. A full-scale implementation of an earned value management system is a complex undertaking. For ease of explanation we will keep the descriptions simple.

To conduct an earned value analysis you need to have planned the work so that each work package [the lowest level of the work breakdown structure (WBS)] has cost and duration estimates. The work is then scheduled and the costs are allocated to the time period in which the work is scheduled. The information below shows the work packages for landscaping a backyard. The work packages are on the left, the duration (in days) is shown in the shaded areas, and the costs associated with the work are indicated in the duration.

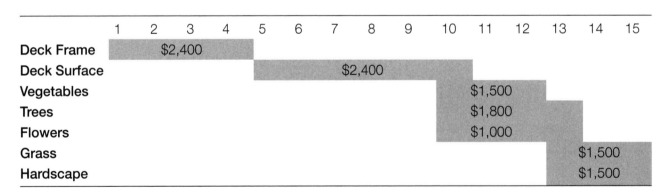

This integration of the approved scope, schedule, and cost for the project is called a performance measurement baseline (PMB). The PMB is used to measure cost and schedule progress. The PMB shows the planned value for each work package at a given point in time. Planned value (PV) is the authorized budget assigned to scheduled work. The planned value for the deck frame at the end of Day 4 is $2,400. Assuming an even distribution of work, the planned value for the deck surface is $800 at the end of Day 6 (assuming that each day $400 of work is scheduled). Thus, the cumulative PV at the end of Day 6 is $3,200.

HOW TO USE IT

Once the PMB is complete you can start measuring progress against it.

1. Identify the value of the work that was accomplished during the work period. This is called the earned value. In the example above, if the deck frame was finished at the end of Day 3, the earned value (EV) at the end of Day 3 would be $2,400. That is because the value of the deck frame is $2,400 and it is complete, therefore you have earned $2,400 worth of value.
2. Analyze the schedule performance by comparing the earned value with the planned value. If you earned less than you planned, you are behind schedule. If you earned more than you planned, you are ahead of schedule. The mathematical equation is EV – PV = Schedule Variance (SV).
3. Determine the amount you spent to accomplish the work. You can do this by looking at invoices, hours spent and the rate per hour, bills of material, and any other costs that went into developing the budget for the work. The amount you spent is called the actual cost (AC).
4. Analyze the cost performance by comparing the EV and AC. If you spent more than you earned, you are over budget. If you spent less than you earned, you are under budget. The mathematical equation is EV – AC = cost variance (CV).

Scenario: Build Twin Pines Medical Plaza, a new state-of-the-art medical resource center.

The Twin Pines Medical Plaza (TPMP) has the following performance measurement baseline for construction.

	Jan	Feb	Mar	Apr	May	Jun	Jul	Aug	Sept	Oct	Nov
Groundbreaking	4,500,000										
Foundation		7,500,000	7,500,000								
Steel Frame				9,000,000	9,000,000						
Shell						13,500,000	13,500,000	13,500,000			
Interior Frame									6,000,000	6,000,000	6,000,000
Monthly Total	4,500,000	7,500,000	7,500,000	9,000,000	9,000,000	13,500,000	13,500,000	13,500,000	6,000,000	6,000,000	6,000,000
Cumulative Total	4,500,000	12,000,000	19,500,000	28,500,000	37,500,000	51,000,000	64,500,000	78,000,000	84,000,000	90,000,000	96,000,000

	Dec	Jan	Feb	Mar	Apr	May	Jun	Jul	Aug	Sept	Oct
Plumbing	3,000,000	3,000,000									
Electric			3,000,000	3,000,000							
HVAC					2,250,000	2,250,000					
Gas							1,500,000				
Alarm								1,500,000	1,500,000		
Telephone								1,500,000	1,500,000		
Cable								1,500,000	1,500,000		
Finish Work										1,500,000	1,500,000
Monthly Total	3,000,000	3,000,000	3,000,000	3,000,000	2,250,000	2,250,000	1,500,000	4,500,000	4,500,000	1,500,000	1,500,000
Cumulative Total	99,000,000	102,000,000	105,000,000	108,000,000	110,250,000	112,500,000	114,000,000	118,500,000	123,000,000	124,500,000	126,000,000

The cumulative total at the end is called the budget at completion (BAC). Thus, the BAC for the construction work is $126,000,000.

Assume that it is the end of August in Year 1. Your PMB shows you have a cumulative planned value of $78,000,000. Your earned value shows you have completed the groundbreaking, the foundation, the steel frame, and 85 percent of the shell. The table below summarizes the results:

Work	Planned Value	Percent Complete	Earned Value
Groundbreaking	4,500,000	100	4,500,000
Foundation	15,000,000	100	15,000,000
Steel Frame	18,000,000	100	18,000,000
Shell	40,500,000	85	34,425,000
Total	**78,000,000**		**71,925,000**

Notice that you haven't indicated expenses for the work done, only the PV and the EV. The actual cost is determined by totaling the invoices to see what has been paid. In this case, the invoices show that the groundbreaking, which is on a fixed price contract, cost the same as the estimate ($4,500,000). The foundation had an overrun of $874,000. The steel frame and shell are also overrunning their cost estimates. The actual costs for the work to date are shown below:

Work	Planned Value	Percent Complete	Earned Value	Actual Costs
Groundbreaking	4,500,000	100	4,500,000	4,500,000
Foundation	15,000,000	100	15,000,000	15,874,000
Steel Frame	18,000,000	100	18,000,000	18,952,250
Shell	40,500,000	85	34,425,000	36,105,800
Total	**78,000,000**		**71,925,000**	**75,504,150**

By looking at the table above, you can intuitively determine that things are not going well. It is obvious that the project is earning less than planned and the cost is greater than the value earned. With earned value techniques, you can make some simple calculations to quantify the results.

We will look at the cost and schedule variances to determine how far behind and how much overbudget the project is running. To measure schedule performance, compare the EV to the PV. To measure cost performance, compare the EV against the AC. We will apply the cumulative PV, EV, and AC information to the equations for cost and schedule variance to come up with the following:

$$EV - PV = SV \qquad 71,925,000 - 78,000,000 = -6,075,000$$

$$EV - AC = CV \qquad 71,925,000 - 75,504,150 = -3,579,150$$

(continued)

(continued)

This tells us that we have accomplished $6,075,000 less work than we planned during the time we spent. We have spent $3,579,150 more than the amount of work we have accomplished. In other words, we are behind schedule and overbudget.

When looking at variances, any negative variance indicates that your performance isn't going how you want. You're either accomplishing less or spending more than you planned. Notice that looking at the earned value numbers doesn't tell you what happened; it only points out that things aren't going according to plan. It is important to determine root cause of the variance in order to take appropriate corrective action to get the project back on plan.

Additional Information

There is much more to earned value analysis. You will see another way to use earned value analysis when we look at performance indexes (Section 2.8), estimate to complete (Section 4.4), estimate at completion (Section 4.3), to-complete performance indexes (Section 4.6), and variance at completion (Section 4.8).

PMBOK® Guide – Sixth Edition References

4.4 Monitor and Control Project Work
6.6 Control Schedule
7.4 Control Costs
12.3 Control Procurements

2.6 INFLUENCE DIAGRAMS

WHAT IT IS

An influence diagram is a graphic that shows the relationship between variables, uncertainty, and the degree of influence of each variable on other variables, decisions, and the ultimate objective. They are used to build a common understanding of the variables that influence decisions and outcomes.

Influence diagrams use various shapes to indicate nodes. The nodes are connected by arrows to show uncertainty, decisions, and objectives. Common shapes for nodes and their meanings are shown below.

An oval node indicates an area of uncertainty.

A rectangular node shows a decision.

An octagonal node (or sometimes a diamond) indicates the objective or desired outcome.

HOW TO USE IT

Use the steps below as a guideline.

1. Identify the objective you want to achieve. This is your octagon.
2. Identify the decisions or choices you will need to make to achieve that objective. These are your rectangles.

3. Determine all the variables that influence that decision or choice. Variables can be things inside your control (availability of resources), or outside your control (such as weather or regulations). These are your ovals.
4. Draw arrows that show the influence of each node on the other nodes.

Scenario: You have been asked to meet the physical growth needs of Top Dog Project Services.

Over the past four years the rate of employee growth and turnover has continued to increase in the Southern California location of Top Dog Project Services. You are looking for ways to accommodate the employee growth and reduce the amount of employee turnover. One of the recurring themes expressed in exit interviews is that employees want the freedom to work remotely.

Top Dog Project Services is looking to run a pilot work-from-home program. Given the needs of your office, they have asked you to run a pilot program for the entire organization for working remotely. They have asked that you summarize the outcomes and make a recommendation to senior management.

As part of your kickoff meeting, you lay out some of the variables that will influence your ability to roll out the program, thereby reducing the rate of turnover.

Influence diagram

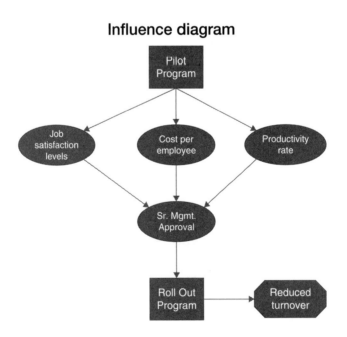

This graph shows the pilot will produce variable results for job satisfaction (either increased or decreased), cost per employee (lower or higher), and productivity rate (lower or higher). Based on these results, senior management will decide to approve a rollout of the program across the company, as one way of reducing employee turnover.

Additional Information

An influence diagram is sometimes used in place of a decision tree because the proliferation of branches based on the nodes of uncertainty can quickly cause a decision tree to get overly complicated and difficult to interpret. The influence diagram also does a better job of showing causality between the various nodes than a decision tree does.

PMBOK® Guide – Sixth Edition References

11.4 Perform Quantitative Risk Analysis

2.7 MAKE-OR-BUY ANALYSIS

WHAT IT IS

A make-or-buy analysis is used to determine which work or deliverables should be accomplished in-house, and which should be purchased from external sources. Factors such as resource availability and skill sets, cost, proprietary information, risks, and other relevant factors are taken into consideration when conducting a make-or-buy analysis.

HOW TO USE IT

In some situations it is obvious you will need to procure something in support of project completion, such as if you don't have the skills available in your organization to develop the deliverable. Sometimes the reasons to keep work in-house are obvious, such as if the deliverable is your organization's core competency, if you need control over the process, or if you want the information to stay in-house.

For those situations that fall in between, here are some factors to consider in your analysis. Use the information below as a guideline. You will need to tailor the information to fit within your environment.

Reasons to Make	Reasons to Buy
Less costly[1]	Less costly[2]
Use existing capacity	Use supplier skills
Maintain direct control	Small volume requirement
Maintain design secrecy	Limited in-house capacity
Develop a new competency	Maintain supplier relations
Company's core competency	Transfer risk
Available staff	Readily available

[1] When analyzing costs, don't forget indirect costs. Indirect costs for making can include overhead, depreciation, tax, material handling and storage, and so forth.
[2] Indirect costs for buying can include the time and cost of developing a Request for Proposal, and the time involved in source evaluation, selection, and management.

Scenario: You are the project manager for developing an onsite childcare center for your organization's employees.

You need to determine whether you will hire people to run the center as staff, or whether you will outsource the staffing and management to Watch Us Grow, an organization that manages childcare centers. You identify the following reasons to hire staff and manage the center in-house:

1. You will save the 15 percent service fee that Watch Us Grow charges.
2. You won't have to expend the time and effort involved with contract negotiation.
3. You will have more control over the curriculum, hours, policies, procedures, and day-to-day operations.

You identify the following reasons to outsource the management of the childcare center:

1. No time or cost associated with job posting, interviewing, checking credentials, background checks, and the like.
2. This is not your core competency. You don't plan to grow this employee benefit into a larger business. You only need five staff to manage the entire program.
3. All risks associated with staff liability are transferred to Watch Us Grow.
4. The time associated with negotiating the contract will be less than the time needed to hire the staff.

The costs associated with the management fee versus the costs associated with managing the regulations, staffing, and other collateral costs are likely close to even. The relative importance of the control over the childcare center operations compared to the reduced time and liability of outsourcing will help determine the decision.

Additional Information

A make-or-buy analysis is often used along with a cost benefit analysis (Section 2.2), net present value calculations, and payback period calculations to help determine the best approach. You can also use a decision tree (Section 2.4) to model different scenarios in a make-or-buy analysis.

PMBOK®Guide – Sixth Edition References

12.1 Plan Procurement Management

2.8 PERFORMANCE INDEX

WHAT IT IS

The performance index uses information from earned value analysis (Section 2.5) to determine the schedule and cost performance efficiency. The schedule performance index (SPI) is a measure of schedule efficiency shown as the ratio of earned value to planned value. It tells us the rate at which we are progressing compared to the planned rate as of a particular point in time. The cost performance index (CPI) is a measure of cost efficiency shown as the ratio of earned value to actual cost. It tells us how much we are getting for every dollar we put into the project as of a particular point in time.

HOW TO USE IT

1. To calculate the SPI, start with the earned value (EV) and divide by the planned value (PV). The equation is EV/PV = SPI.
2. To calculate the CPI, start with the earned value (EV) and divide by the actual cost (AC). The equation is EV/AC = CPI.

Scenario: Build Twin Pines Medical Plaza, a new state-of-the-art medical resource center.

Using the same information introduced in the earned value analysis technique (Section 2.5) you can compute the SPI and CPI. This is a summary of the project performance so far:

Work	Planned Value	Percent Complete	Earned Value	Actual Costs
Groundbreaking	4,500,000	100	4,500,000	4,500,000
Foundation	15,000,000	100	15,000,000	15,874,000
Steel Frame	18,000,000	100	18,000,000	18,952,250
Shell	40,500,000	85	34,425,000	36,105,800
Total	**78,000,000**		**71,925,000**	**75,504,150**

Using the SPI and CPI equations you calculate the following:

$$EV/PV = SPI \qquad 71,925,000/78,000,000 = .92$$

$$EV/AC = CPI \qquad 71,925,000/75,504,150 = .95$$

This information indicates that the project is 92 percent efficient on schedule performance and 95 percent efficient on cost performance. Another way of looking at this is for every 10 days of work, you are getting the value of 9.2 days of work done. For every dollar spent, you are getting .95 cents worth of value.

When working with indexes, any index less than 1.0 indicates that your performance is not going well. You're either accomplishing less or spending more than you had planned.

Additional Information

Performance indexes can be used with estimate to complete (Section 4.4) and estimate at completion (Section 4.3).

The cost performance index is the single best indicator of the overall project health. You may be able to make up schedule variances, but rarely can cost variances be corrected. If there are significant variances at the start of the project, it is likely they will only get worse over time. The reasoning is, if you are having trouble estimating and accomplishing the immediate horizon in work, you will have more trouble calculating the further-out work.

Rather than finishing 75 percent of the project and discovering you are going to run out of funds before completion, using performance indexes will provide early warning signals to allow for corrective action.

Performance indexes are used to identify those areas that need the most attention. The lower the performance index, the more trouble the deliverable or project is in.

PMBOK® Guide – Sixth Edition References

4.4 Monitor and Control Project Work
6.6 Control Schedule
7.4 Control Costs
12.3 Control Procurements

2.9 REGRESSION ANALYSIS

WHAT IT IS

Regression analysis is used to evaluate the relationship between an outcome, known as the dependent variable, and one or more independent variables. Analyzing the relationship between an outcome and one independent variable is called a simple regression model. Analyzing the relationship between an outcome and multiple independent variables is called a multiple regression model.

Regression analysis is used to predict future trends or performance based on the relationship between the dependent and independent variables. It is most often used to predict financial performance, especially for investments. However, it can also be used in healthcare, social sciences, and, on occasion, in projects.

Using regression analysis in projects looks for predictors of dependent variables such as cost performance, schedule performance, or defect rates. Because you need enough data to produce results that are not skewed by having too small of a data set, it is more useful for a PMO to run regression analyses to identify those variables that are the strongest predictors of project success or failure.

Multiple regression analysis requires statistical software to develop a viable model. However, if you are conducting a simple regression analysis you can use an Excel chart and develop a scatter diagram using the two variables to determine if a relationship exists, and if it does, to determine the strength of that relationship.

HOW TO USE IT

The explanation for regression analysis in this book is very limited in scope. We are constraining the description to a simple regression analysis that you can perform using a spreadsheet. This explanation does not assume an understanding of statistical modeling, and therefore we will not describe the method of least squares nor how to determine slope or unexplained residual errors.

1. Define the variable you want to predict. This is called the dependent variable. The dependent variable is always on the *y* axis of a graph.
2. Identify the variable you think will be a reasonable predictor of the dependent variable. The independent variable will be on the *x* axis of a graph.

3. Gather as much data as you can on the dependent and independent variables.
4. Chart them on the graph to create a scatter diagram.
5. Determine if there is a linear relationship.
6. Draw a line through the center scatter diagram that depicts the general trend of the relationship. If you are using spreadsheet software, such as Excel, go to the Chart Design tab and click Add Chart Element. Go to the Trend Line option and select Linear.

Scenario: Your project is to help improve customer satisfaction with the phone support from the IT Help Desk.

Assume you want to know about the variables that drive up the number of complaints for the IT Help Desk. You have an assumption that more seasoned staff get fewer complaints. To test this hypothesis, you determine the number of months each of the ten employees has worked at the Help Desk. The newest hire has been there for eight months. The longest-term employee has been there 64 months. You then look at the number of complaints per employee; they range from 2 to 13 per employee.

You set up a graph with the number of complaints as the dependent variable, and graph it on the *y* axis. The number of months on the Help Desk is the independent variable, and so you graph that on the *x* axis. Using Excel,[1] this is how you set up the data:

Months Employed	8	12	15	26	30	42	45	52	60	64
Complaints	13	10	9	7	6	5	4	3	3	2

Next, you select the cells with data and go to the Insert tab and choose a scatter diagram (see Section 3.10 for more on scatter diagrams). The result is this:

Complaints and Length of Employment

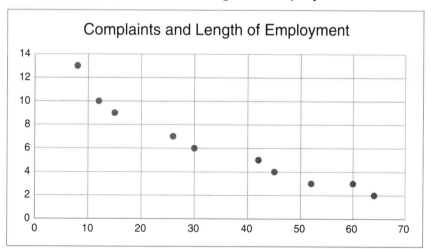

---

[1] Different versions and different operating systems for Excel may have different functions and commands. Your software may have different options and names. The function should be there, but you may have to get to it in a different way.

(continued)

It is clear there is a negative correlation between the length of employment and the number of complaints. In other words, the more experience the employee has, the fewer complaints he or she gets. As a final step, you go to the Chart Design tab and click Add Chart Element. Under the Trend Line option you select Linear. The result is this:

Complaints and Length of Employment

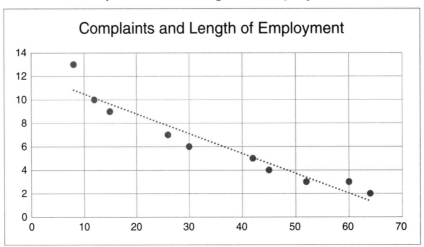

The trend line indicates the expected number of complaints based on the length of employment. Based on this, it looks like the newest employee is still having more complaints than expected, but that could just be an anomaly. You should watch that number for a few months to see if it comes down to the expected level.

Additional Information

There are several points to be aware of when using regression analysis.

- Make sure your data is accurate. Entering inaccurate information will give you skewed or inaccurate results.
- Don't mistake correlation for causation. Just because you have found a correlation between two variables does not mean that one causes the other. In the example above, the lower complaints for the more seasoned employees could be a function of additional training, relationships with the people calling in, having a better office environment, or any number of other variables. The data in the example only indicates that longer-term employees have fewer complaints. The data does not say why this is so.
- When using regression analysis, always check your data with the real world. In other words, don't rely only on the numbers; investigate, talk to people, and look around.
- Remember, data is not a 100 percent predictor of the future; it can be a useful indicator, but it is only a model.

PMBOK® Guide – Sixth Edition Reference

4.7 Close Project or Phase

2.10 RESERVE ANALYSIS

WHAT IT IS

Reserve analysis is used to determine the appropriate amount of schedule or cost reserve necessary to establish an achievable and reasonable cost or schedule baseline. Reserve analysis is conducted prior to developing a baseline, after the initial risk identification, analysis, and response planning have taken place. The purpose is to account for both the individual and overall risks associated with the project and to allocate a suitable amount of time and/or funding to meet the project objectives.

After the baselines are established and the project is underway, you can conduct a reserve analysis to determine if the reserve amounts are sufficient given the most recent project schedule and cost performance. If the project is performing ahead of schedule or under budget, it may be feasible to "release" some of the reserve. If the project is performing behind schedule or over budget it may be appropriate to increase the reserve.

When conducting a reserve analysis you should specify whether you are assessing contingency reserve or management reserve. Contingency reserve is usually set aside for known risk events or conditions. It can be used to implement a response strategy if a risk trigger occurs, or it can be applied to lessen the residual risk impacts. Contingency reserve is usually included in the project baselines.

Management reserve is usually held at the project level to account for unplanned, in-scope work, or for unforeseen events. Management reserve is not usually included in the project baselines. Management reserve may be held at the program or portfolio level if the project is part of a program or portfolio. Some projects only use contingency reserve; others use both contingency and management reserve. The use of management reserve is most often employed at the discretion of management and is typically not considered a part of the approved cost baseline.

The cost estimates plus the contingency reserve comprise the cost baseline. The cost baseline plus the management reserve equal the project budget. The graphic below shows how different types of reserve are integrated into the project budget.

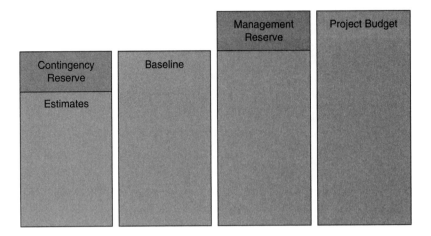

Reserve can be estimated as a percent of the overall duration or budget, as a fixed number, or it may be developed using a simulation or other quantitative analysis method.

HOW TO USE IT

The example below discusses a budget reserve. The same steps are used to develop a schedule reserve. Use the steps below as a guideline. Tailor the steps as necessary to work within your environment.

To develop an initial reserve amount:

1. Develop cost estimates for all the project work.
2. Develop a schedule to show the allocation of project work over time.
3. Create a budget that shows the allocation of project funds over time.
4. Conduct a thorough risk identification, analysis, and response exercise, including establishing a contingency reserve to implement responses and to account for those risks you accept.

The outputs of these first four steps provide you with

- Project Schedule
- Project Budget
- Risk Register that is populated with risk events, probabilities, impacts on each objective, and planned responses.
- Assessment of overall project risk (such as information on ambiguity, variability, volatility, uncertainty, and so forth)

5. Estimate the amount of reserve you will need using one or more of these techniques:
   a. Sum all the contingency reserve amounts for individual risks and choose a percentage of that for the contingency reserve for the project.
   b. Take a set percent of the overall project budget.
   c. If you are using simulation software, run a simulation and select the reserve amount that corresponds with the confidence level you want to achieve.

To assess the risk reserve throughout the project:

1. Identify the amount of reserve remaining.
2. Assess the current budget performance (are you over or under budget, and by how much).
3. Analyze the Risk Register and Risk Reports to determine the amount of risk remaining on the project.
4. Compare the amount of reserve remaining to the amount of risk remaining to determine if the remaining reserve is sufficient.

Scenario: You have been asked to meet the physical growth needs of Top Dog Project Services.

You are working with the Pennsylvania office to lease a building in a new office park to accommodate the expansion of Top Dog's workforce. The work is broken into the following phases:

- Organize
- Design
- Build
- Move

During the organizing phase, you have developed preliminary cost estimates and a budget. Below is the budget summary by category:

Project Management	$56,250
Design	$225,000
Build	$731,250
Move	$22,500
Total	$1,035,000

Your risk identification and analysis indicates the following risk exposure as a percent of the budget by phase:

Phase	Percent of Category Budget	Dollar Amount
Project Management	5	$2,250
Design	11	$24,750
Build	8	$58,500
Move	Flat amount	$4,500
Total		$90,000

The project seems fairly low risk, but you want to put in some contingency reserve to bring up the amount to 10 percent of the budget. So you add $13,000 in management reserve to account for unknown risks.

Additional Information

When establishing reserve for the schedule it is a good practice to include some reserve prior to phase endings and key deliveries to ensure these milestones are met.

PMBOK® Guide – Sixth Edition References

6.4 Estimate Durations
7.2 Estimate Costs
7.3 Determine Budget
7.4 Control Costs
11.7 Monitor Risks

2.11 ROOT CAUSE ANALYSIS

WHAT IT IS

Root cause analysis, often abbreviated as RCA, is a technique used to identify the underlying cause of a variance, risk, or defect. Often when a problem arises we rush to fix it by treating the symptom, for example, crashing the schedule to get back on track. Root cause analysis looks beneath the symptoms to find out what underlying events, conditions, systems, or processes allowed the event or condition to happen in the first place.

The underlying assumption when conducting a root cause analysis is that variances, defects, and risks have an underlying cause in a system or process. The goal of RCA is to trace the problem back to its source so that you can fix the source of the problem. Most defects or variances can be attributed to one of three categories:

1. **Physical causes.** Physical causes include faulty material or equipment, or environmental issues such as poor lighting, bad weather, confusing instructions, and the like.
2. **Human causes.** Human causes are based on someone doing something he or she shouldn't have, or not doing something he or she should have.
3. **Organizational causes.** Organizational causes are policies, procedures, systems, or processes that govern work, or help people make decisions.

HOW TO USE IT

Use the steps below as a guideline. Tailor the steps as necessary to work within your environment.

1. Define the variance, defect, or problem. Sometimes being able to clearly articulate the problem can be challenging, so spending time coming up with a defined problem statement is important. Other times, it is fairly obvious, such as "we are 7 percent over budget" or "the deliverable is not functioning the way it should."
2. Collect all the facts and data around the variance, defect, or risk. For complex problems, this can take some time and perseverance to understand the full situation. When collecting information about the problem, it is productive to get many different perspectives on what is happening to gain a full understanding of the situation.

3. Identify contributing factors. These are events, conditions, or circumstances that increase the likelihood of the variance or problem occurring. You can use the "5 whys" scenario, where you keep asking "why" until you reach the root of the problem. Another technique is to keep asking "so what" to determine all the possible consequences of a fact. Cause-and-effect diagrams are a good way to get a visual representation of all the contributing factors.

4. Analyze the contributing factors to determine the underlying assumption, process, event, or system that allowed the variance, problem, or risk to occur. Once you have boiled down the contributing factors, you will find one or more root causes that need to be addressed to fix the problem.

5. Recommend and implement solutions. Once the root cause is identified, you will need to figure out the appropriate steps to take to ensure it does not happen again. You will need to assess the risks associated with various solutions, assign someone to follow through, and check back to see if the solution fixed the problem.

Scenario: Your project is to help improve customer satisfaction with the phone support from the IT Help Desk.

The research you did with check sheets (Section 1.3) showed six categories of complaints. You decide to conduct a root cause analysis on the category with the most complaints first—"being on hold too long." Twenty-two of sixty-two total complaints are in this category (35 percent). In this case, the problem is easy to define:

Problem Statement: Customers are unhappy with the amount of time they spend on hold when they call the Help Desk.

To understand the situation, you get all the ACD reports that show the average hold time, the hold time by time of day, hold time by day of the week, and the average hold time for each Help Desk employee. You also talk to some of the people who made the complaints and the people at the Help Desk to get their feedback on the situation. You walk around the Help Desk area to get a sense of the environment and you review the Help Desk Manual.

While you are interviewing the Help Desk employees, you ask under what circumstances they place a caller on hold. You use a version of the 5 whys until you fully understand the various reasons why calls are put on hold. You start to draw a cause-and-effect diagram (see 3.1 for more information on cause-and-effect diagrams) to identify the major sources of putting calls on hold.

Based on assessing all this information, you identify several sources of lengthy hold times:

1. The Help Desk added two new employees in the past year; however, there was not enough room to give them full cubes. They have partial cubes and sit next to a break room. There are times when there is a lot of background noise or conversations going on around them. The Help Desk employees put the calls on hold, wait for a break in the conversation, and then ask the people having the conversation to please keep their volume down so the employee can focus on the call.

2. Some of the systems the Help Desk is supporting are older "legacy" systems. The seasoned employees have experience working with these systems and have built up their knowledge

over the years. These systems are not documented online. All the documentation is paper based. Often, a Help Desk employee needs to reference a binder to find the information he or she needs to help the caller.
3. There is very little training on the systems for new Help Desk employees. Much of the learning happens as they go.

Having identified these three root causes for extended hold times, you make the following recommendations:

1. Find a quieter place for the Help Desk team.
2. Develop more robust training for Help Desk employees.
3. Partner new employees with a seasoned employee for mentoring.
4. When calls about legacy systems come in, refer them to a seasoned employee, and have the new employee listen in to learn the system.

Additional Information

With some projects you are just trying to figure out what caused a deliverable to be late, and the process will be relatively simple and quick. Some process improvement projects are based on a root cause analysis and implementing the proposed solution. The steps in this process can be scaled to fit your needs.

You will need to apply some common sense with regard to how long you keep tracing things back to the root cause. You want to balance stopping when you find a root cause, with the risk that there may be additional causes, with continuing on well past the point where it is useful.

PMBOK® Guide – Sixth Edition References

4.5 Monitor and Control Project Work
8.2 Manage Quality
8.3 Control Quality
11.2 Identify Risks
13.2 Plan Stakeholder Management
13.4 Monitor Stakeholder Engagement

2.12 SENSITIVITY ANALYSIS

WHAT IT IS

A sensitivity analysis is a modeling technique used to determine how different values of an input impact the values of specific outputs. Sensitivity analysis is used in many different industries for many different purposes. For example, sensitivity models are used to model climate change, investment decisions, economic forecasts, and so on. Most projects don't use big, robust sensitivity analysis models. For the most part, on projects we are looking to identify the variables that have the biggest impact on cost variability and schedule variability. Therefore, you will likely see it used in larger projects when developing schedules and budgets and quantifying risk.

There are different outputs from a sensitivity analysis. Projects that conduct sensitivity analyses will often use a tornado diagram, so this section will present information with the assumption that the preferred output is a tornado diagram.

There are several potential uses for a sensitivity analysis. Among them:

* Identify the variables that result in the greatest schedule variation.
* Identify the variables that result in the greatest cost variation.
* Identify the variables that result in the greatest technical performance variation.
* Identify the variables that result in the greatest defect variation.
* Gain a greater understanding of the relationship between environmental conditions or inputs, and results or outcomes.

As mentioned above, there are many ways to develop a sensitivity analysis; the focus here is on a simple method that looks at systematically assessing the impact of variation in one factor at a time (known as OFAT in the world of sensitivity analysis), while keeping all other variables stable. By conducting an OFAT analysis, you will be able to identify those factors that create the most variation in the outcome (aka the most uncertainty), and identify ways to reduce or eliminate the uncertainty.

HOW TO USE IT

Use the steps below as a guideline. Tailor the steps as necessary to work within your environment.

1. Select the output you want to study (schedule, costs, technical performance, defects, and so on).
2. Identify the drivers (aka inputs) that you think have the most impact on the variable.

3. Determine the high and low values for each input.
    a. You can use the extremes, or you can use a range of 10 to 90 percent. In other words, in normal circumstances you would expect the outcome to be between the 10 and 90 percent limits.
    b. Some models also include the expected value (aka base or nominal value).
4. Determine the likely outcome of the output based on the variation of one selected input, holding all other variables stable.
5. Return the first input to its nominal state, select the next variable, and repeat the process until all variables have been evaluated.
6. Graph the results using horizontal bars that show those variables that have the greatest degree of variation from the nominal value on the output at the top of the chart, and those with the lowest impact on the bottom of the chart. The outcome is shaped a bit like a tornado, hence the name tornado diagram.

Scenario: Your organization is updating its current PMO information system infrastructure with all new software, cloud computing, collaboration sites, and real-time reporting software.

The expectation is that you will market this platform for a monthly subscription fee to use all the features. The project will take one year to complete, so the income will not be realized until next year.

The decision to launch this capability was based on several assumptions. You have been asked to determine which assumptions have the greatest impact on the projected first-year profit. Thus, the profit is the output you want to study.

Assumptions

1. 40 percent of existing 12,500 clients will add the new service in the first year.
2. The company will add another 40 percent more in sales to new customers in the first year.
3. The subscription rate is $250 per month.
4. The interest rate is 5 percent.
5. Maintenance costs are $45,000 per year.

Based on these assumptions the gross sales would be $2,500,000, with a net profit of $2,338,095.

$$\big((12{,}500 \times 40\%) + (12{,}500 \times 40\%)\big) \times \$250 = \$2{,}500{,}000$$

$$\big(\$2{,}500{,}000 - \$45{,}000\big)/1.05 = \$2{,}338{,}095$$

You believe these assumptions are the largest drivers of variability for the profit. If these assumptions are not valid, you may not meet the target profit.

You use the assumptions as the baseline. From there you interview sales and marketing, financial analysts, and the IT department to derive the lowest and highest expected values for each variable.

(continued)

(continued)

We will assume for this example you are only using a spreadsheet without any macros or fancy calculations. Your initial table showing the assumption and the low value, high value, and baseline value is below.

	Low	High	Baseline
Adoption rate existing	2,500	7,500	5,000
New clients	3,750	6,250	5,000
Subscription rate	200	350	250
Interest rate	1.04	1.06	1.05
Maintenance costs	40,000	60,000	45,000

Note the far-right column contains all the baseline assumptions.

Your first iteration holds the value for new clients, subscription rate, interest rate, and maintenance at their baseline. Then you calculate the gross sales and net profit assuming a higher adoption rate and a lower adoption rate. The results are below.

Adoption Rate from Existing Clients

	Gross Sales	Net Profit
Base	$2,500,000.00	$2,338,095.24
High	$3,125,000.00	$3,082,142.86
Low	$1,875,000.00	$1,742,857.14

You return the adoption rate to the baseline number. Next you calculate the gross sales and net profit assuming more sales from new clients and less sales from new clients. You return the new sales to the baseline number and repeat the process with the subscription rate, interest rate, and maintenance costs. After running all those calculations you end up with the following ranges in outcomes for each variable.

	High	Low
Adoption rate existing	$3,082,143	$1,742,857
New clients	$2,635,714	$2,040,476
Subscription rate	$3,290,476	$1,861,905
Interest rate	$2,316,038	$2,360,577
Maintenance costs	$2,323,810	$2,342,857

This is good information, but to make it easier to analyze you start with the baseline net profit and find the difference (variance) between each outcome.

	Positive	Negative
Adoption rate existing	$744,048	$(595,238)
New clients	$297,619	$(297,619)
Subscription rate	$952,381	$(476,190)
Interest rate	$22,482	$(22,057)
Maintenance costs	$4,762	$(14,285)

This information is what you turn into a tornado diagram. First you sort the data from low to high. Then you insert a bar chart. The end result is the following diagram:

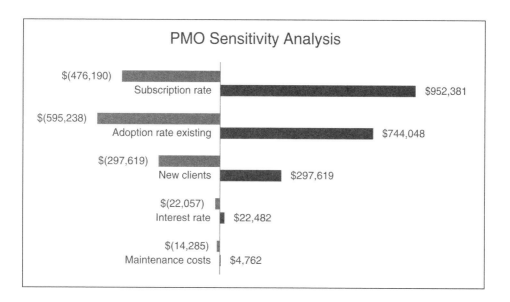

It is obvious that the biggest driver of profit is the subscription price, followed by the adoption of rate of existing clients and then the acquisition of new subscribers. The interest rate and maintenance costs have minimal impact.

This sensitivity analysis does not show the interaction that the impact of raising or lowering the subscription rate would have on adoption rates or new subscriptions. More sophisticated software is needed to conduct an analysis where multiple variables are correlated.

Additional Information

A sensitivity analysis is listed in the *PMBOK® Guide* as a technique to identify those activities or cost items that have the most influence on the end result. The length of the bar indicates the relative influence. The longer the bar, the more influence.

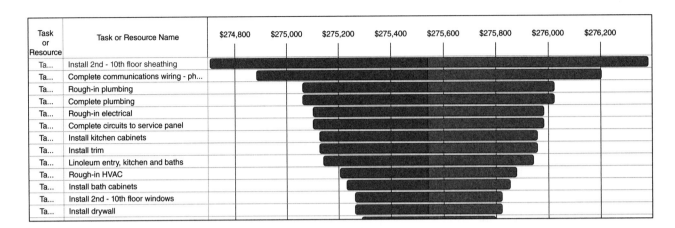

Task or Resource	Task or Resource Name	$274,800	$275,000	$275,200	$275,400	$275,600	$275,800	$276,000	$276,200
Ta...	Install 2nd - 10th floor sheathing								
Ta...	Complete communications wiring - ph...								
Ta...	Rough-in plumbing								
Ta...	Complete plumbing								
Ta...	Rough-in electrical								
Ta...	Complete circuits to service panel								
Ta...	Install kitchen cabinets								
Ta...	Install trim								
Ta...	Linoleum entry, kitchen and baths								
Ta...	Rough-in HVAC								
Ta...	Install bath cabinets								
Ta...	Install 2nd - 10th floor windows								
Ta...	Install drywall								

This screenshot shows an excerpt on cost sensitivity for project activities. The bars on the right show the possible amount of a budget overrun; and those on the left, the possible amount of budget underrun.

The information derived from a sensitivity analysis can be used in working with a what-if scenario (see Section 2.17).

There are several issues to keep in mind when working with a sensitivity analysis:

- Often the variables are correlated, which means that it is unrealistic to think that you can change one variable and all others will remain stable. The interactions among the variables means that only looking at one variable at a time can lead to unrealistic results. There is software that can model more complex models that show the interaction of multiple variables. Unless you are working on a mega-project, it is unlikely that you will need to employ that degree of rigor in your analysis.
- If assumptions that are used to determine the high and low values for the variables are based on past data, the data may not be reliable for the current situation.
- Estimates are inherently subjective, thus the data for the model may not be factual, but subject to human bias.

PMBOK® Guide – Sixth Edition References

11.4 Perform Quantitative Risk Analysis

2.13 STAKEHOLDER ANALYSIS

WHAT IT IS

Stakeholder analysis is used to gather and assess information about stakeholders, the people or groups that can influence or are influenced by the project outcomes, to determine how to best engage with them. This technique is used throughout the project as stakeholders and their engagement needs change and evolve. Stakeholder analysis is the underlying foundation for an iterative five-step process used in stakeholder communication and engagement:[1]

1. Identify your project's stakeholders and understand their needs.
2. Prioritize the stakeholders.
3. Map the key stakeholders.
4. Engage with the stakeholders.
5. Monitor changes over time to assess the effectiveness of your engagement.

HOW TO USE IT

Use the steps below as a guideline. Tailor the steps as necessary to work within your environment.

1. Identify stakeholders and gather information about them. Information you can assess in a stakeholder analysis includes:[2]
   ○ Attitude: Will the person help or hinder the work?
   ○ Hierarchy: Where is the person in the organization's structure compared to the activity manager: higher or lower, internal or external, colleague or competitor?

---

[1] Based on work by Lynda Bourne and Patrick Weaver, Mosaic Projects. https://stakeholder-management.com/
[2] Bourne, L. (2015). *Making Projects Work: Effective Stakeholder and Communication Management.* Boca Raton, FL: Young and Francis.

○ Interest: Does the person have an active interest, passive interest, or no interest?
○ Power: What is the person's ability to cause change on the project?
○ Proximity: How involved is the person in the work?
○ Urgency: Does the person perceive the work to be very important?

For projects that have a complex and influential stakeholder community, you may want to collect all this information. For smaller projects, you can focus on the nature of each stakeholder's influence (upward, downward, outward, and sideways), and his or her interest in the project.

2. Prioritize your stakeholders. There are various models you can use to prioritize your stakeholders. One method is by assessing the power, proximity, and urgency of each stakeholder.
   a. Power is the ability to change or stop the project.
   b. Proximity is the degree of involvement a stakeholder has in the project.
   c. Urgency is the relative importance of the project and its outcomes to a stakeholder and his or her business needs.
3. Map the data. It is a good practice to develop a graphic or tabular representation of the stakeholder information to help illustrate the stakeholder landscape. This is a tabular representation based on power, proximity, and urgency.

Stakeholder	Power	Proximity	Urgency
Stakeholder 1	High	High	Medium
Stakeholder 2	Medium	Low	Medium
Stakeholder 3	Medium	High	High
Stakeholder 4	Low	Medium	Medium

Another methodology uses a stakeholder cube[3] to map data using the categories of attitude, power, and interest.

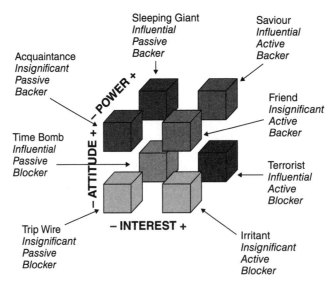

The analysis is used to help develop a stakeholder engagement strategy and to monitor the effectiveness of the strategy throughout the life of the project.

---

[3] Hillson, D. A., and Simon, P. W. (2012). "Practical Project Risk Management: The ATOM Methodology" (2nd edition), Management Concepts.

Scenario: You are the project manager for a project to implement a childcare facility for your organization's employees.

At the very beginning of the project you start a stakeholder analysis by identifying the following stakeholders:

- Sponsor
- Facilities department
- Project manager
- Project team
- Parents
- Children
- Contractors
- Company employees
- Childcare staff

Based on your knowledge of each stakeholder and on interviews and discussions with all parties, you determine each individual's power over the project, his or her proximity (degree of involvement), and his or her urgency (how important the project is).

You then create a table to illustrate the stakeholder landscape.

Stakeholder	Power	Proximity	Urgency
Sponsor	High	Medium	Medium
Facilities department	Medium	High	Medium
Project manager	Medium	High	High
Project team	Low	High	High
Parents	High	Medium	High
Children	Low	High	High
Contractors	Low	High	Low
Company employees	Low	Low	Low
Childcare staff	Medium	Low	High

This information will guide how you communicate and engage with each stakeholder and stakeholder group. As new stakeholders arise, you will add them to your matrix and update your analysis. If your engagement strategy is not working as well as you would like it to, you may evaluate your analysis to see if you assessed someone's involvement incorrectly.

Additional Information

For a more robust stakeholder analysis you can consider these ways of categorizing stakeholders as well:[4]

- Receptiveness: How easy is it to communicate with this person?
- Supportiveness: Does the person support or oppose the work?
- Influence: How well connected is the person?
- Legitimacy: Does the person have some level of entitlement to be consulted?

PMBOK® Guide – Sixth Edition References

11.1 Plan Risk Management
13.1 Identify Stakeholders
13.4 Monitor Stakeholder Engagement

---

[4] Bourne, L. (2015). *Making Projects Work: Effective Stakeholder and Communication Management.* Boca Raton, FL: Young and Francis.

2.14 SWOT ANALYSIS

WHAT IT IS

SWOT stands for Strengths, Weaknesses, Opportunities, and Threats. It is often used in product development projects to help build a business case. It can also be used for a risk assessment prior to authorizing a project or during the project.

A SWOT analysis looks at internal and external aspects of an organization.

* Internal factors are the organization's strengths and weaknesses.
* External factors are opportunities and threats for the organization.

For purposes of a SWOT analysis we will consider the following:

* Strengths are those aspects of the organization that give it an advantage, or those things the firm does well.
* Weaknesses are those aspects that put an organization at a disadvantage in the marketplace, or those areas that require improvement.
* Opportunities are aspects in the external environment that an organization can take advantage of or exploit to improve its business position.
* Threats are aspects of the external environment that could impede or negatively impact the organization's business position.

Some areas to look at when identifying strengths and weaknesses include:

* **Human resources.** Do you have the skill sets and availability you need?
* **Physical resources.** Do you have the right location, equipment, supplies, and materials?
* **Finances.** Do you have an adequate funding stream?
* **Systems.** Are your systems and processes adequate?
* **Expertise.** Do you have the experience and expertise required?
* **Intellectual capital and property.** Do you have copyrights, patents, and other intellectual property advantages or disadvantages?

Some areas to look at when identifying opportunities and threats include:

- **Political environment.** Is the political environment favorable for what you are trying to achieve? Is it stable or unstable? Friendly or unfriendly?
- **Economy.** Is the economy in the markets you are in healthy or declining?
- **Exchange rate.** If you are working on international projects, is the exchange rate in your favor?
- **Trends in your industry.** Is your industry growing or declining?
- **Location.** Does your location give you an advantage or a disadvantage?
- **Legislation and regulations.** Are there legal requirements that constrain your options or enable you to excel?

HOW TO USE IT

A SWOT analysis usually requires input from multiple sources since the objective is to get a broad understanding of both the internal and external environments. You can use many of the other techniques described in this book to help you, such as brainstorming (Section 1.2), focus groups (Section 1.5), and mind mapping (Section 3.6). Use the steps below as a guideline. Tailor the steps as necessary to work within your environment.

1. Identify the strengths and weaknesses internal to the project or organization.
2. Identify the opportunities and threats external to the organization.
3. Assess how the strengths can be turned into opportunities or used to minimize the threats.
4. Determine how the weaknesses can hinder opportunities or increase threats.

Scenario: You have been asked to meet the physical growth needs of Top Dog Project Services.

Based on the success of the work-from-home pilot, you are expanding the program. Before kicking off the program you want to understand the aspects of the organization that can help make the program a success and those that can hinder it. You also want to determine if there are any opportunities that this program will allow you to capture, or if there are new threats it would introduce. This information will be documented in the Risk Register so that you can develop appropriate responses as part of the program implementation.

You gather a small group of people from HR, IT, Compliance, and Operations. You ask them to focus first on the following topics for strengths and weaknesses before brainstorming additional areas.

- Human resources
- Physical resources
- Systems, processes, and policies

Once the group has generated all the information they can think of, you move to looking at external threats and opportunities. You ask the group to focus on the following topics for opportunities and threats before brainstorming additional areas.

- Job market
- Industry trends
- Legislation and regulations

Once the group has generated all the information they can think of, you thank the participants for their assistance. You will use this information to populate your Risk Register and to identify overall risks and opportunities associated with the program.

Additional Information

Many projects neglect opportunity management. Identifying internal strengths is a good way to leverage opportunities for both your project and the organization.

PMBOK® Guide – Sixth Edition Reference

11.2 Identify Risks

2.15 TECHNICAL PERFORMANCE ANALYSIS

WHAT IT IS

Technical performance analysis compares planned technical achievement to actual technical achievement. This analysis technique is often used in system engineering, information technology, and defense projects. The technique requires you to identify capabilities that have quantifiable measurements as targets, so that you can measure progress against those targets.

Examples of capabilities or requirements include, but are not limited to:

- Weight
- Size
- Transaction time
- Errors or defects
- Availability
- Capacity

HOW TO USE IT

Use the steps below as a guideline. Tailor the steps as necessary to work within your environment.

1. Identify the capabilities you need to meet the project objectives.
2. Determine the technical and operational specifications or metrics needed to achieve the capabilities.
3. Determine when you need to achieve each technical and operational specification or metric.
4. Define periodic measures of performance you use to track progress against the requirements.
5. Take measures.
6. Determine if any variances represent a threat or opportunity to the project.

Scenario: You are managing a project to develop a new company intranet site.

The technical capabilities and measurements you have defined are:

Capability	Measure
Site hits per minute	2,500
Load time per page	.001 seconds
Availability	99%

You decide to build up the ability to accept more site hits on a weekly basis. Every week you should have the capacity to receive more site hits per minute without crashing. You set your measurement time as every Friday afternoon at 4 PM. This chart shows progress against the requirement.

Site Hits Per Minute

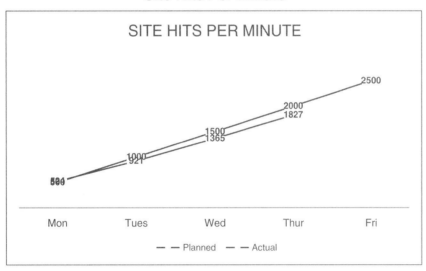

Based on this information you can see that your technical performance is behind. Likely you will be around 2,300 by the end of Week 5. You will need to determine if that performance variance presents a risk to the scope or schedule objectives of the project, and if so, determine an appropriate response.

Additional Information

This technique is often used with burnup or burndown charts to show performance against expectations. When used in risk monitoring, one should determine if the technical performance variance will lead to a scope, schedule, or cost risk.

***PMBOK® Guide* – Sixth Edition Reference**

11.7 Monitor Risks

2.16 VARIANCE ANALYSIS

WHAT IT IS

Variance analysis measures the amount of variance between a planned outcome (often a baseline) and actual performance.

You can use variance analysis to measure variance in:

- Schedule dates
- Expenditures
- Resource utilization
- Quality metrics
- Technical performance

Basically, if you can measure it, you can perform a variance analysis.

HOW TO USE IT

Use the steps below as a guideline. Tailor the steps as necessary to work within your environment.

1. Establish your planned or target measure (such as a date or dollar amount).
2. Measure your actual result.
3. Identify the cause of the variance (if it isn't already obvious).
4. Determine if the variance warrants preventive or corrective action.

Scenario: You have been asked to meet the physical growth needs of Top Dog Project Services.

You work for the Idaho Falls location of Top Dog Project Services. You are remodeling an old downtown building as part of an urban renewal program. Part of the building remodel is to put in steel framing in the interior. You are tracking the cost and schedule performance for the steel frame.

You budgeted $200,000 for the frame. The work is now complete. You review your invoices for labor and materials and you see that the frame actually cost $206,032. In looking at your invoices for the labor, you see that the rates were higher than estimated and the hours were greater than estimated. The estimated rates and hours and the actual rates and hours are below:

Estimated hourly rate $28
Actual hourly rate $30.25
Estimated hours   960
Actual hours   1,088

To understand the impact of the rate variance you subtract the actual rate from the estimated rate, and multiply that by the actual hours worked.

$$(\text{Estimated rate} - \text{Actual rate}) \times \text{Actual hours}$$

$$(\$28 - \$30.25) \times 1088 = \$ - 2,448$$

To understand the impact of the variance in the hours worked, you subtract the actual hours worked from the estimated hours worked, and multiply that by the estimated rate.

$$(\text{Estimated hours} - \text{Actual hours}) \times \$28$$

$$(960 - 1,088) \times \$28 = \$ -3,584$$

Therefore, you determine that the $2,448 of our variance is due to a difference in the hourly rate and $3,584 is due to the difference in the number of hours worked. At this point the steel framing is complete. You have enough reserve to cover the $6,032 variance, so no further action is required.

Additional Information

You can calculate the same information with materials variance as well. You can compare the difference in the cost per unit and in the number of units used. This is sometimes called a cost/usage variance.

Variance analysis is also used in earned value management. The three most common earned value variance measurements are:

Schedule variance (SV): $SV = EV - PV$
Cost variance (CV): $CV = EV - AC$
Variance at completion (VAC): $VAC = BAC - EAC$

PMBOK® Guide – Sixth Edition References

4.5 Monitor and Control Project Work
4.7 Close Project or Phase
5.6 Control Scope
6.6 Control Schedule
7.4 Control Cost

2.17 WHAT-IF ANALYSIS

WHAT IT IS

What-if scenario analysis evaluates various scenarios, risk events, or conditions to determine the impact on a stated objective. It is used primarily to determine the impact of events or conditions on the project schedule or budget.

Some uses of the what-if scenario for the schedule are:

- Evaluating the impact of adding, subtracting, or changing resources
- Evaluating the impact of scope changes
- Assessing the feasibility of the existing schedule if a risk event or condition occurs
- Identifying the amount of schedule reserve needed to account for uncertainty
- Evaluating the impact of fast-tracking the work
- Evaluating the impact of reducing scope

Some of the uses of a what-if scenario for the budget are:

- Evaluating the impact of a change in labor or material rates
- Evaluating the impact of a change in labor or material amounts
- Evaluating the impact of scope changes
- Identifying the impact of currency fluctuations

What-if scenario analysis is also used in risk management to either understand the impact of risk events or to help determine the amount of reserve needed for the schedule or budget.

HOW TO USE IT

Use the steps below as a guideline. Tailor the steps as necessary to work within your environment.

1. Start with a copy of your baseline schedule or budget.
   a. You can use additional columns in a schedule to enter fields such as Start 1, Start 2, and so forth to show different start dates instead of using a copy of the schedule.
   b. Many spreadsheets give you the ability to identify specific fields to enter in alternative numbers. This is part of the software's "what-if analysis" function that allows you to model scenarios.

2. Identify and document the scenario you want to model, including the corresponding assumptions and basis of estimates if you have them.
3. Make the changes to the schedule or budget to reflect your documented scenario.
4. Compare the scenario to the baseline.
   a. For the schedule, you can determine the impact on the critical path, resource utilization, start and end dates, and durations.
   b. For the budget, you can evaluate the impact on labor costs, material costs, funding requirements, or other variables.

Modeling Techniques

So much uncertainty is associated with projects that it's nearly impossible to determine actual durations, future risks, stakeholder issues, actual resource availability, and so on. One way to account for the uncertainty is to think through different scenarios that could impact the project and see what happens to the schedule based on those scenarios.

For example, what if a critical resource is available only one-half of the time, as opposed to full time? Or, what if a key component is delivered late? Or design issues cause rework? These are common occurrences on projects. Building several schedule scenarios helps you to identify the impact of these scenarios up front when you can still adjust the schedule to account for them, and update the Risk Register to take actions to minimize their impact on the schedule.

Scenario: Your organization is updating its current PMO information system infrastructure with all new software, cloud computing, collaboration sites, and real-time reporting software.

The expectation is that you will market this platform for a monthly subscription fee to use all the features. The project will take one year to complete, so the income will not be realized until next year.

The decision to launch this capability was based on the following assumptions.

Assumptions:
1. Forty percent of the existing 12,500 clients will add the new service in the first year.
2. The company will add another 40 percent due to new sales in the first year.
3. The subscription rate is $250 per month.
4. The interest rate is 5 percent.
5. Maintenance costs are $45,000 per year.

Based on these assumptions, the gross sales for Year 2 would be $2,500,000, with a net profit of $2,338,095.

$$\big((12{,}500 \times 40\%) + (12{,}500 \times 40\%)\big) \times \$250 = \$2{,}500{,}000$$

$$\big(\$2{,}500{,}000 - \$45{,}000\big)/1.05 = \$2{,}338{,}095$$

(continued)

You have been asked to look at the impact to the profit if the existing client adoption and new client acquisition are less than expected. You decide to run two sets of budget scenarios:

1. Option 1 assumes 20, 25, 30, and 35 percent adoption rates for existing clients, and maintains the 40 percent for new sales.
2. Option 2 assumes 20, 25, 30, and 35 percent adoption rates for both existing clients and new sales.

You think the range of outcomes will provide sufficient information for financial risk analysis based on the adoption and new sales rates. Your calculations give you the following information:

Adoption Rate	Existing Clients	Existing Clients and New Customers
40%	$2,338,095	$2,338,095
35%	$2,189,286	$2,040,476
30%	$2,083,333	$1,785,714
25%	$1,891,667	$1,445,238
20%	$1,832,143	$1,207,143

You chart the data to see the information graphically to model expected revenue based on a specific adoption rate.

What-If Scenario Analysis

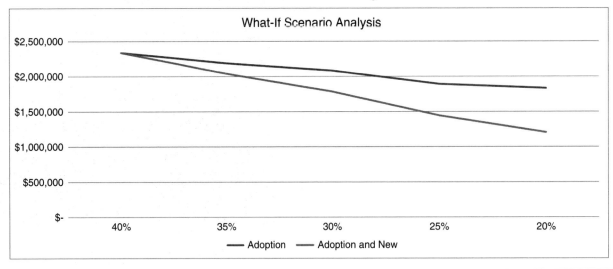

Additional Information

What-if scenarios are similar to sensitivity analyses (Section 2.12), but with a what-if scenario analysis you can model the impacts of more than one variable at a time. You can also develop scenarios based on a single event, rather than a range of outcomes.

PMBOK® Guide – Sixth Edition References

6.5 Develop Schedule
6.6 Control Schedule

Part 3

Data Representation

3.0 DATA REPRESENTATION TECHNIQUES

To really understand what data is telling you, sometimes you need to see it in a graphic representation. As they say, a picture is worth a thousand words. The techniques in this section are used to help create a visual display of data to aid in organizing and understanding what is happening on your project.

Many of these techniques have their beginnings in quality management; for example, control charts, cause-and-effect diagrams, and histograms. Some types of charts we use to help us organize information, such as resource breakdown structures or responsibility assignment matrixes. Other techniques are used to help solve problems, such as cause-and-effect diagrams.

The techniques described in this section include:

- Cause and effect diagrams
- Control charts
- Flowcharts
- Histograms
- Logical data models
- Mind mapping
- Probability and impact matrix
- Resource breakdown structure
- Responsibility assignment matrix
- Scatter diagrams
- Stakeholder mapping

Some of the techniques you will learn about for data representation are general in nature, such as flowcharts and mind mapping. You will see other techniques that are used in very specialized ways, such as a probability and impact matrix or stakeholder mapping.

When using data representation techniques, you may find it easier to just grab a pencil and paper and start mapping the situation. For example, a flowchart can be very informal. Other techniques work better with software. For example, you can use a spreadsheet or other charting software to create a histogram or scatter diagram.

When summarizing information for stakeholders in a presentation, I find it useful to include charts and graphs, such as the ones in this section, to help drive home a point, or to help make the information come to life.

3.1 CAUSE-AND-EFFECT DIAGRAM

WHAT IT IS

A cause-and-effect diagram illustrates the inputs or variables that are potential causes of risks or defects. By working backward from the defect, you ask: How did that happen? or Why did that happen? These questions lead to the inputs to the process or the product that could be potential causes of defects. You can break each branch of the diagram into limbs and continue to decompose them until you have sufficient detail to identify the source of the problem.

The figure here shows a classic cause-and-effect diagram that lists six possible causes of a problem: environment, measurement, energy, personnel, equipment, and time. Two of the categories have subcategories.

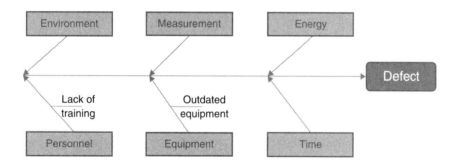

Cause-and-effect diagrams can be relatively simple, like the one shown here; or they can be more complex, with many branches and interrelationships between the causes.

HOW TO USE IT

You can conduct a cause-and-effect analysis with a group, as shown below, or on your own. Working with a small, knowledgeable group usually leads to better outcomes.

1. Define the problem statement (this is the outcome or defect you are trying to resolve).
2. Brainstorm the major categories of the causes of the problem. If you are having trouble identifying categories, you can use the ones shown in the example above.

3. For each category, ask "Why does this happen?"
4. Each answer to why is a subcategory that is represented as a branch.
5. Continue to ask why and develop further subcategories until the diagram is complete.

Scenario: Your project is to help improve customer satisfaction with the phone support from the IT Help Desk.

One of the main causes of dissatisfaction is that callers feel they are placed on hold for too long. You call a meeting with the Help Desk staff to brainstorm the factors that lead to calls being put on hold. You come up with the following cause-and-effect diagram.

Cause-and-effect diagram

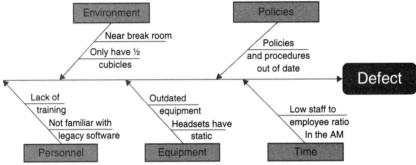

The notes that support the diagram are captured below:

- **Time:** The time of day can affect the number of people staffing the Help Desk and the number of people at work.
- **Policies:** There are few policies and procedures for the Help Desk, and those that are available are out of date.
- **Equipment:** The headsets that the employees use are outdated, and sometimes cause static. The legacy systems have no electronic training material, only outdated paper manuals.
- **Measurements:** The measurements for the automatic call distribution system (ACD) metrics are measuring the call start time from the time a person picks up the phone, not from the time the call goes into the ACD queue.
- **Personnel:** There are two relatively new employees and there is no formal training that takes place. Even though staff members have the technical skills, they're not familiar with some of the legacy software.
- **Environment:** The Help Desk staff don't have full cubicles, and they sit near the break room, where people congregate and chat. There is a lot of background noise.

Additional Information

A cause-and-effect diagram is also known as a *fishbone diagram* or an *Ishikawa diagram.*

The cause-and-effect diagram can be used to illustrate causes and potential causes when conducting a root cause analysis (Section 2.11).

PMBOK® Guide – Sixth Edition References

8.2 Manage Quality
8.3 Control Quality

3.2 CONTROL CHARTS

WHAT IT IS

A control chart is a graphic display of a process over time. Data points are plotted and compared to a central line (the average) and upper and lower control limits. Control charts are more common in a production or factory environment, where they can be used to measure process results. However, they can be used to measure project results, such as defects, cost and schedule variance, or any other predictable event on a project.

For example, it is predictable that you will have cost and schedule variances on your project. However, you want to make sure they are within an allowable limit (tolerance limit). You can track the variances using a control chart to spot trends and to monitor the cost and schedule performance.

Control charts measure the performance of a process to ensure it is in control. When a process is in control the results of the process are consistent with expectations. When it is not in control there are outliers or unexpected results. Control charts can be used in quality planning, when establishing processes and tolerance levels that determine whether the process is in control.

When using control charts for statistical process analysis, the upper and lower control limits are +/- 3 standard deviations from the mean. Specification limits represent an acceptable variance from the mean. This means that any result that falls between the upper and lower specification limits are acceptable; any result that is outside those limits is not acceptable. Specification limits may be greater than or less than the area defined by the control limits. The upper and lower specification limits are defined in the quality requirements or quality metrics. If a measurement is getting close to the control limit or specification limit, you should take action to get it back toward the midline.

HOW TO USE IT

Use the steps below as a guideline. Tailor the steps as necessary to work within your environment.

1. When planning for product or project quality, establish the items that you will measure with a control chart.
2. Set the upper and lower control limits and/or upper and lower specification limits. If you are not doing statistical process analysis, you do not need to set the upper and lower control limits at +/- 3 standard deviations from the mean. Set the upper and lower specification limits to reflect the variance tolerance for whatever you are measuring.

3. In a spreadsheet column, enter the time intervals you will use to measure the process, such as daily, hourly, weekly, and so forth.
4. In the next column, enter the measurement data.

Here are two columns that show measurement data for the cost performance index (see Section 2.8 for information on the cost performance index, CPI) measured weekly over the second quarter.

Weekly	CPI
4/1	0.97
4/8	0.98
4/15	0.99
4/22	0.96
4/29	0.95
5/6	0.93
5/13	0.91
5/20	0.88
5/27	0.89
6/3	0.92
6/10	0.94
6/17	0.95
6/24	0.95

5. In the column next to the measurement data (in this case, next to the CPI column), enter the average or target. If you are doing a statistical process analysis you will use the average measurement. For our purposes we will enter the target value. For CPI the target value is 1.0.
6. In the column next to the average or target value, enter your upper control limit (UCL) and your lower control limit (LCL). The chart will now look like this:

Weekly	CPI	Target	UCL	LCL
4/1	0.97	1.00	1.1	0.9
4/8	0.98	1.00	1.1	0.9
4/15	0.99	1.00	1.1	0.9
4/22	0.96	1.00	1.1	0.9
4/29	0.95	1.00	1.1	0.9
5/6	0.93	1.00	1.1	0.9
5/13	0.91	1.00	1.1	0.9
5/20	0.88	1.00	1.1	0.9
5/27	0.89	1.00	1.1	0.9
6/3	0.92	1.00	1.1	0.9
6/10	0.94	1.00	1.1	0.9
6/17	0.95	1.00	1.1	0.9
6/24	0.95	1.00	1.1	0.9

7. If you are going to track performance against upper and lower specification limits (USL and LSL), enter those next to the upper and lower control limits.

Weekly	CPI	Target	UCL	LCL	USL	LSL
4/1	0.97	1.00	1.1	0.9	1.05	0.95
4/8	0.98	1.00	1.1	0.9	1.05	0.95
4/15	0.99	1.00	1.1	0.9	1.05	0.95
4/22	0.96	1.00	1.1	0.9	1.05	0.95
4/29	0.95	1.00	1.1	0.9	1.05	0.95
5/6	0.93	1.00	1.1	0.9	1.05	0.95
5/13	0.91	1.00	1.1	0.9	1.05	0.95
5/20	0.88	1.00	1.1	0.9	1.05	0.95
5/27	0.89	1.00	1.1	0.9	1.05	0.95
6/3	0.92	1.00	1.1	0.9	1.05	0.95
6/10	0.94	1.00	1.1	0.9	1.05	0.95
6/17	0.95	1.00	1.1	0.9	1.05	0.95
6/24	0.95	1.00	1.1	0.9	1.05	0.95

8. Highlight all the data. Go to the Insert Tab and select a Line Chart. The result should look like this:

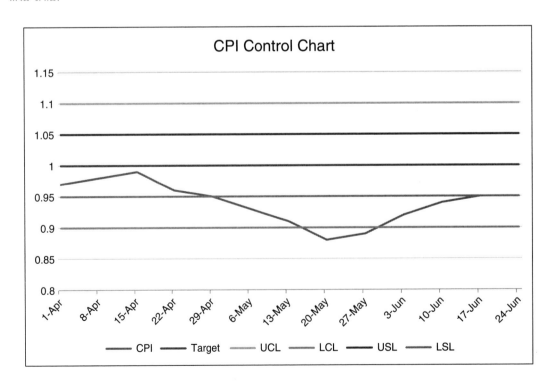

Scenario: Your project is to help improve customer satisfaction with the phone support from the IT Help Desk.

One of the problems with the Help Desk is that it takes a long time for someone to answer the call. You decide to track the amount of time it takes to get a live person on the line over the course of the day. The Help Desk is staffed from 6:00 AM to 8:00 PM. The ACD system collects information on the average number of seconds before calls are picked up.

The current policy states that calls should be answered by a person (not a machine) within 60 seconds. Therefore, you set your target as 60 seconds. You set an upper specification limit of 90 seconds and a lower specification of 5 seconds. This is what your table looks like:

Hour	Seconds	Target	Lower Spec Limit	Upper Spec Limit
6:00 AM	15	60	5	90
7:00 AM	23	60	5	90
8:00 AM	63	60	5	90
9:00 AM	88	60	5	90
10:00 AM	108	60	5	90
11:00 AM	104	60	5	90
12:00 PM	88	60	5	90
1:00 PM	113	60	5	90
2:00 PM	82	60	5	90
3:00 PM	72	60	5	90
4:00 PM	54	60	5	90
5:00 PM	23	60	5	90
6:00 PM	40	60	5	90
7:00 PM	18	60	5	90
8:00 PM	9	60	5	90

You insert a line chart and see this:

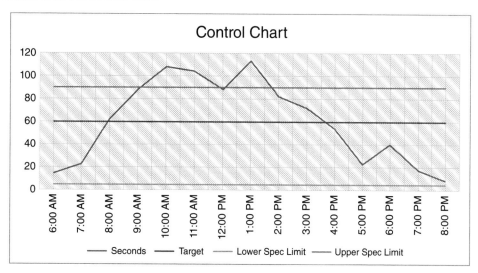

(continued)

(continued)

From 10:00 AM to noon and from 1:00 to 2:00 PM, the average hold time is greater than the upper specification. You can use this information to assist with staffing levels during the middle of the day, or develop other resolutions.

Additional Information

You might hear about the Rule of Seven when talking about control charts. The Rule of Seven states that seven data points trending in one direction (up or down) or seven data points on one side of the mean indicate that the process isn't random. This means that you should check the measurement to determine whether something is influencing the process. This would be a special cause variation because it's highly unlikely that in a random process you would find this type of behavior.

PMBOK® Guide – Sixth Edition Reference

8.3 Control Quality

3.3 FLOWCHARTS

WHAT IT IS

Flowcharts are used to develop a model of a process by identifying all the inputs, actions, outputs, and decision points in a process. You can use them to define, standardize, and communicate a process.

Flowcharts are often used in process improvement projects. By reviewing flowcharts of a process you can identify bottlenecks and redundancies and eliminate steps that provide no real value.

HOW TO USE IT

You can map all the steps by hand on a piece of paper. Another useful technique is to use sticky notes, where each note is a step in the process. You can also use flowchart software.

1. Identify all the steps in a process.
2. Put the tasks in order by asking: "What has to happen first?" or "After this step, what do I do?"
3. Create a flowchart of the sequence of tasks linked by arrows.

Scenario: Your project is to help improve customer satisfaction with the phone support from the IT Help Desk.

One of the first steps in identifying the source of problems is to develop a flowchart of the process that the caller goes through when calling the Help Desk. The outcome is this flowchart, which shows the series of steps a caller goes through when contacting the Help Desk.

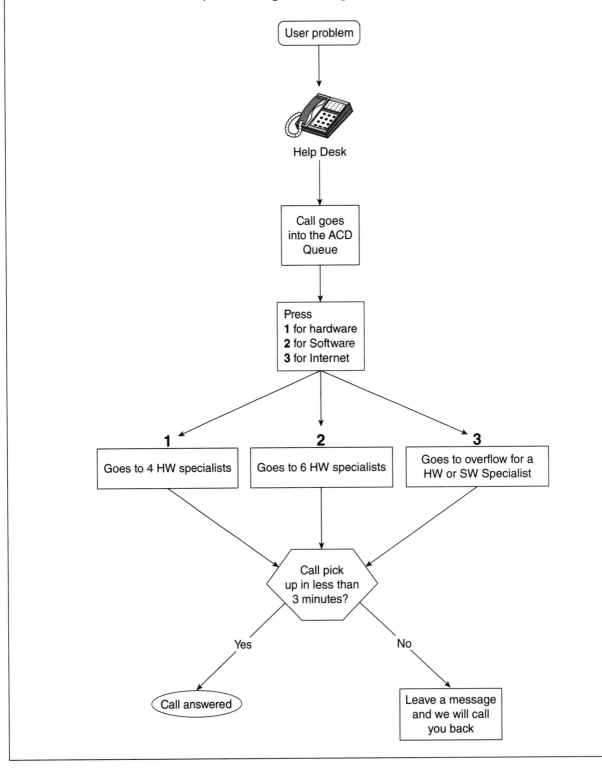

Additional Information

Sometimes flowcharts are called process maps. They are often used in process improvement projects to document the "As Is" process and the "To Be" process.

There are some standards associated with flowchart shapes. For example:

An elongated oval marks the start or finish of a process

Rectangles represent activities

Diamonds indicate a decision or choice

There are many different shapes and icons that represent different steps and functions in a process. For complicated processes, it is easier to work with software designed to do flowcharts and process mapping.

PMBOK® Guide – Sixth Edition References

8.1 Plan Quality Management
8.3 Control Quality

3.4 HISTOGRAMS

WHAT IT IS

A histogram is a bar chart that shows the frequency of various events. In project management it is usually used to support the quality management processes by showing the distribution of errors or defects.

If you arrange a histogram in descending order from the most defects to the least, it is known as a Pareto chart or Pareto diagram.

HOW TO USE IT

Since a histogram is used to illustrate the number of times a defect or error occurs, we assume the work of gathering the information is complete. This explanation only explains how to create the chart.

1. In a spreadsheet row, enter the categories you are measuring:

Did not meet spec	Too slow	Overweight	Crashed	Error screen

2. In the row below, enter the number of times there was an occurrence in each category:

Did not meet spec	Too slow	Overweight	Crashed	Error screen
5	17	14	6	12

3. Select the cells with the data, and select Insert Chart.

4. Choose a bar chart (or column chart).

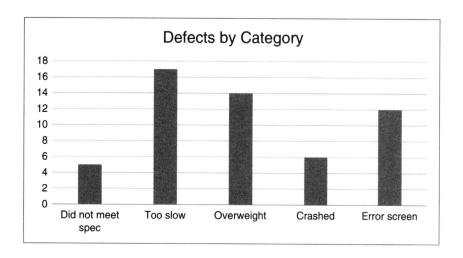

If you want to dress it up a bit, you can add data labels. To add data labels, right click on one of the bars. When a side menu shows up, select Add Data Labels.

To turn it into a Pareto chart, just arrange the numbers from highest to lowest.

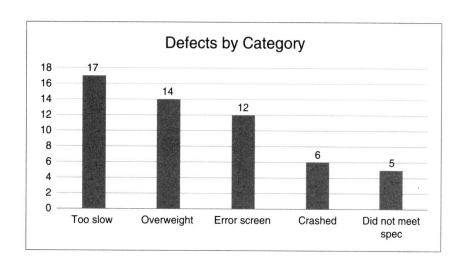

Scenario: Your project is to help improve customer satisfaction with the phone support from the IT Help Desk.

As part of the project to improve satisfaction with the IT Help Desk, the team reviewed the check sheets (see Section 1.3) for the IT department from the employee satisfaction survey for the internal support services. You created a histogram by entering the number of times each category of complaint was indicated on the check sheets. You then ranked these in order, to create a Pareto chart.

(continued)

(continued)

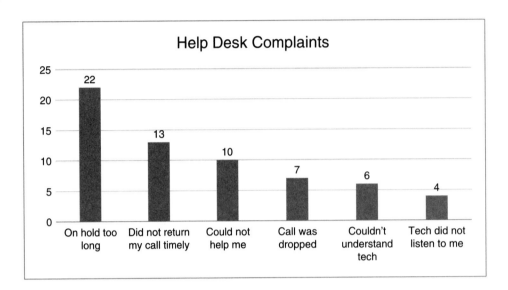

Help Desk Complaints

Based on the histogram, it is clear you need to address the issue of people being on hold too long, and then look at improving the time for calling people back. If you can improve those two issues, more than half of the complaints will be addressed.

Additional Information

A Pareto chart is sometimes referenced in conjunction with the "80/20 rule," which states 80 percent of the problems come from 20 percent of the causes.

PMBOK® Guide – Sixth Edition References

8.2 Manage Quality
8.3 Control Quality

3.5 LOGICAL DATA MODEL

WHAT IT IS

Logical data models are used to show entities, relationships, data, and how data is arranged. Logical data models are used to help promote a common understanding of business data elements and requirements, as well as provide a foundation for designing a database. The data model is used for a particular problem domain and is expressed independently of a specific database product. Logical data models are used in IT projects to provide more detail than a conceptual data model (which shows entities and relationships only). Data models are often diagrammatic and are used to capture information important to an organization. They are technology independent and are strictly about the entities, data, and relationships.

Some of the terms you will need to be familiar with to understand this technique include:

Entity. An object in a data model. For example, an address, a customer, or a product.

Entity (or primary) key. The information that will identify and differentiate one set of data in an entity from any other. For example, if you are identifying a customer you can use the social security number as an entity key. Each customer will have a unique social security number, so that is the entity key that differentiates one customer from another.

Entity attributes. The types of data that will go into the entity. For example, for a product entity you would have a product number, product description, wholesale cost, retail price, and product category.

Entity relationships. Entities can have different relationships such as 1-to-1, 1-to-many, or many-to-many.

HOW TO USE IT

Use the steps below as a guideline. Tailor the steps as necessary to work within your environment.

1. Identify the entities you will include in your model.
2. Map the relationships between the entities.
3. Identify the primary key (or entity key) for each entity.
4. Specify the attributes for each entity.
5. Normalize the data (reduce or eliminate any redundancies).

Scenario: You are the project manager for a project to implement a childcare facility for your organization's employees.

As part of the Childcare Center you are developing a mobile app to check children in and out of the center. This will allow you to know how many and which children are onsite at any given point in time. You will also have information about who can pick up the child, special dietary or medical needs, and their specific learning plans.

The entities you need are:

- Child information
- Family and guardian information
- Developmental information
- Nutrition information
- School readiness

After meeting with the school administrator, you identify the following relationships and primary keys.

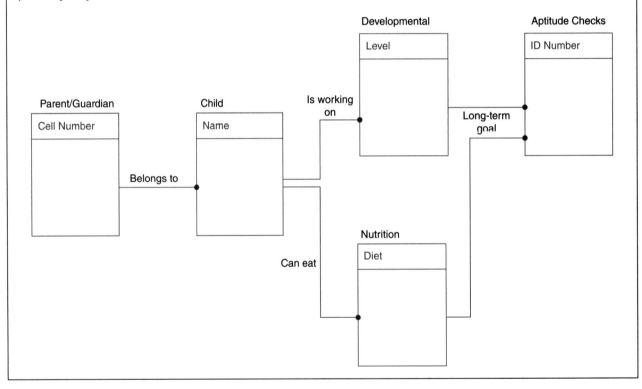

The primary key is shown under the entity. For example, the Parent/Guardian is the entity and the primary key is the cell number. Once you understand how data will flow and be used you can determine the entity attributes. This leaves you with the following completed logical data model:

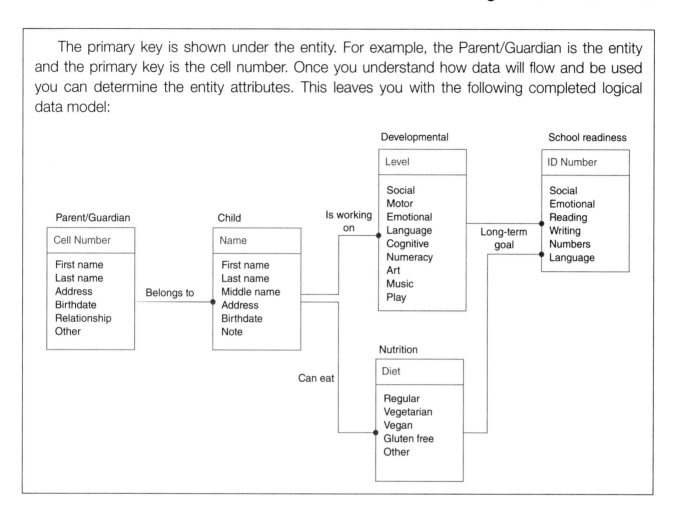

Additional Information

The logical data model becomes an input to developing the physical data model, which you will use to convert the information from the entities into data tables, foreign keys, and data columns. Some examples of logical data models are Entity Relationship Diagram (ERD) or Universal Markup Language (UML).

PMBOK® Guide – Sixth Edition Reference

8.1 Plan Quality Management

3.6 MIND MAPPING

WHAT IT IS

A mind map is a technique that starts with an idea (or project) in the center. Then you branch out from the project and identify the main themes or deliverables associated with the idea. Each theme or deliverable has further details. The "branching out" continues until you have a completed mind map that provides an overview of the entire project or elements of a project under analysis.

HOW TO USE IT

These steps can be used to draw out a mind map manually, or to create one on mind mapping software.

1. Start with the central idea.
2. Develop the major branches or categories associated with the main idea.
3. Fill in the branches with the details (twigs) about the idea.
4. Rearrange and fill in as needed until the resulting mind map is complete.

Scenario: You are the project manager for a project to implement a childcare facility for your organization's employees.

Once the charter for the childcare center is signed, you meet with your core team to brainstorm the components of the center. You start with the childcare center in the middle. Then you decide to look at deliverables associated with the curriculum, the arts, indoor activities, and the playground. You work with your team to break each of these into more detail and end up with a mind map that looks like this:

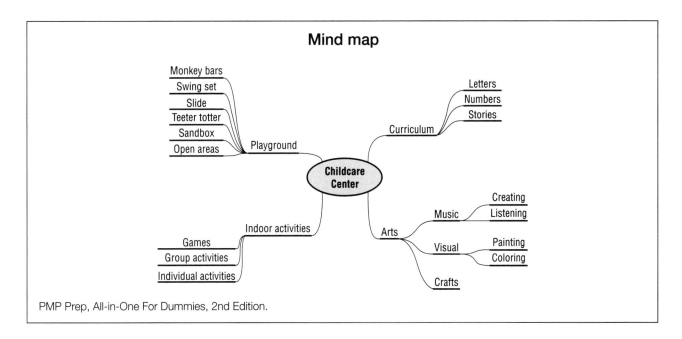

Mind map

Monkey bars
Swing set
Slide
Teeter totter
Sandbox
Open areas
Playground

Curriculum
Letters
Numbers
Stories

Childcare Center

Indoor activities
Games
Group activities
Individual activities

Arts
Music
Creating
Listening
Visual
Painting
Coloring
Crafts

PMP Prep, All-in-One For Dummies, 2nd Edition.

Additional Information

There is software that is designed specifically for mind mapping. You can just as easily use sticky notes so that you can move them around until you settle on a final mind map. If you want to be creative, you can draw, use icons, work with different color markers, and paste pictures on your mind map. Mind mapping can be a very creative and fun way to fill out the project deliverables or requirements.

PMBOK® Guide – Sixth Edition References

5.2 Collect Requirements
8.1 Plan Quality Management
13.2 Plan Stakeholder Engagement

3.7 PROBABILITY AND IMPACT MATRIX

WHAT IT IS

A probability and impact matrix (PxI matrix) is a grid that maps the likelihood of an event occurring and, if it does occur, its impact on project objectives. It is used during risk analysis to help prioritize project risks based on the combination of probability and impact.

When assessing probability you are assessing the likelihood of an event occurring. You are not assessing the likelihood of the event causing an impact, only the likelihood of it occurring. When you assess impact you are assuming the event will occur, and if it does, looking at the impact on one or more objectives.

For smaller or less critical projects, a probability and impact matrix can assess an impact on any objective (for example, scope, quality, cost, schedule, or stakeholder satisfaction). For a complex or large or critical project, you may want to have a separate PxI matrix for each objective.

HOW TO USE IT

Use the steps below as a guideline. Tailor the steps as necessary for your specific project.

1. Begin with a Risk Register that identifies risk events associated with a project.
2. Meet with team members, subject matter experts, or others with an understanding of the risk events. You might also consider any historical data from similar projects and their associated risk documentation.
   a. Use the expertise of those present along with data from previous projects to determine the likelihood of each event occurring.
   b. Use the expertise of those present along with data from previous projects to determine the impact on any project objective if the event does occur.
3. The risk management plan has established definitions of "probability" and "impact." Use the information in the risk management plan to rate the risk according to the definitions in the risk management plan.
   a. For example, if a risk is considered 40 percent likely to occur, and the risk management plan shows a definition of "medium probability" is 40 to 60 percent, you would indicate that the risk has a Medium probability.

b. If the impact to the schedule would be a nine-day slip on the critical path, and the risk management plan indicates that a slip between one and two weeks on the critical path is a High impact, then you would rate the impact as High.

4. If the risk management plan does not already have a grid in place that reflects the probability and impact outcomes, create one.

5. Map the individual risks onto the probability and impact matrix.

Scenario: You are the project manager for a project to implement a childcare facility for your organization's employees.

You call a risk meeting with your project team, the general contractor, and the procurement specialist for the project. You have identified many events or conditions that could occur which would negatively impact the project. We will focus on three that impact the schedule for this example.

Risk 1: Because the security cameras you selected are on back order, there is a possibility that they will not arrive as scheduled. The cameras are an integral part of the security system and need to be wired and tested before the childcare center can open. They are not currently on the critical path; however, if they are more than three weeks late, they will cause a delay to the opening.

Risk 2: The city might not approve plans, causing rework and a delay in the construction phase.

Risk 3: Lily, (a team member) may be reassigned to a higher-priority project, causing a schedule delay while we find a replacement and orient them to the project.

Below is an excerpt from the risk management plan.

DEFINITIONS OF PROBABILITY

Very High—80 to 99 percent likelihood of occurrence
High—60 to 79 percent likelihood of occurrence
Medium—40 to 59 percent likelihood of occurrence
Low—20 to 39 percent likelihood of occurrence
Very Low—1 to 19 percent likelihood of occurrence

DEFINITIONS OF SCHEDULE IMPACT

Very High—A slip on the critical path of greater than two weeks
High—A slip on the critical path greater than one week and less than two weeks
Medium—A slip on the critical path of up to one week
Low—A slip on a noncritical path that uses all the float
Very Low—A slip on a noncritical path that uses some float, but not all the float

Using these definitions, you rate the risks as follows:

ID	Risk	Prob.	Impact
1	Because the security cameras you selected are on back order, there is a possibility that they will not arrive as scheduled.	5	5
2	The city might not approve plans, causing rework and a delay in the construction phase.	2	5
3	Lily may be reassigned to a higher priority project, causing a schedule delay while we find a replacement and orient the new recruit to the project.	2	2

(continued)

(continued)

You then map these risks onto the probability and impact matrix as shown here.

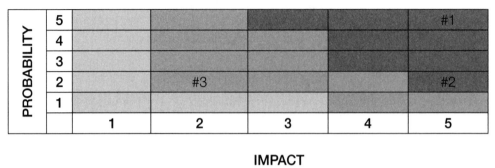

IMPACT

■ = High Risk   ▨ = Medium Risk   ▫ = Low Risk

This ranking shows you which risk events you should address first. Weigh this information with the relative urgency (how close the event is to occurring) to help you fine-tune your priorities.

Additional Information

There are many ways you can tailor your PxI matrix to reflect the risk appetite of your organization. Here are three ways to help you adapt the PxI matrix to be more reflective of your particular project:

1. You can change the definitions of probability and impact, making them more restrictive or looser depending on whether your project requires a more risk-averse or more risk-taking attitude.
2. You can change the numbering scheme for the probability and/or impact. For example, you can change the numbers from 1, 3, 5, 7, 9 to .5, 1, 2, 4, 8 to show a lower value for lower risks. The second scheme doubles the value every time it goes up an increment.
3. You can change the shading of the risk matrix. Putting more squares in the High Risk category reflects a lower risk appetite. Putting more squares in the Low Risk category reflects a greater risk appetite.

PMBOK® Guide – Sixth Edition Reference

11.3 Perform Qualitative Risk Analysis

3.8 RESOURCE BREAKDOWN STRUCTURE

WHAT IT IS

A resource breakdown structure decomposes resources from a high level to a more detailed level. It can be used for physical resources, team resources, or both. You can use a resource breakdown structure for resource planning. At the start of the project, you would have the types of resources; as you progressively elaborate the information, you would get to the numbers of resources, the skill levels of the participants, and ultimately the names of the resources.

HOW TO USE IT

1. Identify the high-level types of resources you will need for your project, such as people, materials, equipment, locations, supplies, and so forth.
2. Decompose each high-level resource into finer levels of detail as appropriate.

Scenario: You have been asked to meet the physical growth needs of Top Dog Project Services.

You are reviewing information for the Idaho Falls location of Top Dog Project Services. They are remodeling an old downtown building as part of an urban renewal program. To help you develop the budget, they are identifying the skills and skill levels they will need to build out the building that is being renovated. This is an excerpt that focuses on the tradespeople who will be working on the project.

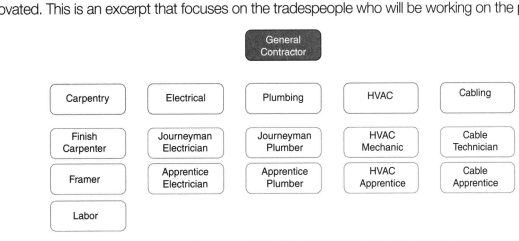

Additional Information

You can use the resource breakdown structure to help with resource optimization (see Section 6.9), for cost estimating, developing the schedule, resource planning, communication planning, risk management, and stakeholder engagement.

PMBOK® Guide – Sixth Edition Reference

9.1 Plan Resource Management

3.9 RESPONSIBILITY ASSIGNMENT MATRIX

WHAT IT IS

A responsibility assignment matrix (RAM) is a matrix with the project work shown in the left-most column and the available resources across the top. The matrix shows the resources assigned to the work, and the way they will support the work.

Project work can be shown at various levels of detail depending on where it is in the project life cycle. At the start of planning the project, work may be shown at a control account or work package level. If you are using progressive elaboration, such as rolling wave planning (Section 6.10), you can show work at the activity level. Resources are also progressively elaborated. At the start of planning the project, you may have a high-level description, such as "programmer," "crane operator," or "engineer." By the time you have detailed plans developed you will likely have the names of the resources that are assigned to the work.

You can categorize the various roles that resources will fulfill to support the work. For example, you might have someone accountable for the work getting done, someone else might sign off on the deliverable, and another person might provide input or consultation for the work. Documenting the type of support helps to communicate expectations for everyone involved in working on or supporting the project.

HOW TO USE IT

Use the steps below as a guideline. Tailor the steps as necessary depending on your environment and where you are in the project life cycle.

1. List each deliverable (or work package, control account, or activity, depending on where you are in the project life cycle) in a column on the left-hand side of a matrix chart.
2. List each resource that has any kind of role in the project across the top of the chart. Your resources may include people who are not part of your team, such as a permitting agency, senior management, outside consultants, and the like.
3. Determine the categories of work or support that will be used on the matrix (responsible for the work, approves the work, etc.).
4. Document the categories in a legend or key so people reading the chart will know how to interpret it.
5. Fill in the matrix to show which resources perform work or support each deliverable.

Scenario: You have been asked to meet the physical growth needs of Top Dog Project Services.

You work for the Idaho Falls location of Top Dog Project Services. You are remodeling an old downtown building as part of an urban renewal program. At this point in the project you are doing high-level planning for the buildout. This is an excerpt from the responsibility assignment matrix (RAM) that focuses on the trades work.

	Project Manager	General Contractor	Electrician	Plumber	HVAC	Cable Technician	City Inspector
Develop Schedule	A	C/S	C/I	C/I	C/I	C/I	
Develop electric schematics	I	S	A				
Develop plumbing schematics	I	S		A			
Develop HVAC schematics	I	S			A		
Develop cabling schematics	I	S				A	
Electrical work	I	S	A				S
Plumbing work	I	S		A			S
HVAC work	I	S			A		S
Cable work	I	S				A	S

Legend:
Accountable
Inform
Sign Off
Consult

This RAM indicates that the tradespeople will be consulted about the schedule, and will be informed when it is developed by the project manager. The general contractor will sign off on all schematics and all finished work. The city inspector will sign off on all finished work. The project manager will be kept informed as each deliverable is completed. Each tradesperson is accountable for performing his or her specific area of work.

When this work is further elaborated there will be more detailed activities, and more detailed information about the resources.

Additional Information

A common type of responsibility assignment matrix is a RACI[1] chart. RACI stands for

R = Responsible. The person who does the work
A = Accountable. The person who ensures the work is done as required. This person may delegate the work to the responsible person, or do the work him- or herself. The accountable party is the person the project manager considers as owning the work.

[1] *A Guide to the Project Management Body of Knowledge*, Sixth Edition. PMI. Newtown Square, PA.

C = Consult. Consult can mean a consultant, or anyone who is consulted for the work. This can include a subject matter expert, a team member, the customer, or any other resource who provides information and input to the deliverable.

I = Inform. These are people who are informed about the work.

In the RACI version there is always one "A" and only one "A." This ensures that there is always only one person who is answerable for the work getting done.

PMBOK® *Guide* – Sixth Edition Reference

9.1 Plan Resource Management

3.10 SCATTER DIAGRAMS

WHAT IT IS

A scatter diagram is a graph plotted along an x and y axis. The purpose is to show the relationship between two variables. For example, if you were to plot the amount of budget spent on a task and the remaining duration of the task, you would expect to see a negative correlation. The duration should shrink as more of the budget is spent. You would expect to see a positive correlation if you were tracking resources added to the project and budget expended.

HOW TO USE IT

A scatter diagram is used to show the relationship between two variables. Thus, we assume the work of gathering the information is complete. This explanation only describes how to create the chart.

1. Identify the two variables you want to assess for correlation.
2. Assign one variable to the x axis and one to the y axis.
3. Plot your data.
4. Select the data and insert a scatter chart.

Scenario: Your project is to help improve customer satisfaction with the phone support from the IT Help Desk.

As part of developing a plan to improve satisfaction with the IT Help Desk you decide to create a few scatter charts to identify some of the key correlations between satisfaction ratings and other variables. The first scatter diagram you develop shows the relationship between the number of Help Desk staff and the percent of time the Help Desk was within the upper specification limit (see Control Charts in Section 3.2 for more information on specification limits). As expected, there was a positive correlation, which means that as one variable increases, the other increases. A perfect 1:1 correlation gives you a straight diagonal line.

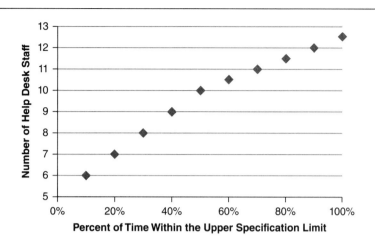

The next thing you measure is the relationship between the number of complaints and the length of employment. This result showed a negative correlation, where the longer someone worked there, the fewer complaints they received.

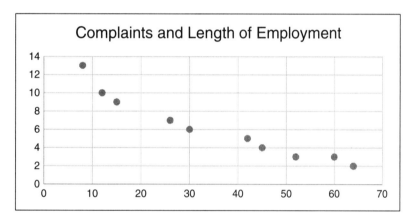

Based on this information you can make some recommendations for changes with regard to staffing and training for newer employees.

Additional Information

Scatter diagrams are the foundation for regression analysis, which is a widely used statistical technique that estimates the relationship among variables. Regression analysis is described in Section 2.9.

In addition to a positive and negative correlation, you may see a curvilinear correlation that starts high, diminishes, and then increases again at the other extreme.

PMBOK® Guide – Sixth Edition References

8.2 Manage Quality
8.3 Control Quality

3.11 STAKEHOLDER MAPPING

WHAT IT IS

A stakeholder engagement assessment matrix is used to show where stakeholders are, and where you would like them to be in relation to their support of the project. Once you have done an initial analysis of the stakeholder community (see Section 2.13), you can take it a step further by creating a stakeholder engagement assessment matrix to help guide your plan to optimize stakeholder engagement.

This matrix shows different levels of awareness and support.

Stakeholder	Unaware	Unsupportive	Neutral	Supportive	Driver*

*The difference between someone who is "supportive" and someone who is a "driver" is the amount of power that person has to impact the direction of the project or to create major change with the project.

HOW TO USE IT

Use the steps below as a guideline. Tailor the steps as necessary to work within your environment.

1. Start with your stakeholder register and your stakeholder analysis results (Section 2.13).
2. Work with your team to review all stakeholders, and plot each one on the chart with a "C" for current state.
3. Discuss where you would like each to be, and what is reasonable to expect. We would all like all stakeholders to be supportive of our projects, but that is not a reasonable expectation.
4. Mark a "D" on the chart to indicate the desired state.

Scenario: You are the project manager for a project to implement a childcare facility for your organization's employees.

You are meeting with your team to analyze your stakeholders. You go through the register to assess the current level of engagement with the stakeholders and come away with the following:

Stakeholder	Unaware	Unsupportive	Neutral	Supportive	Driver
Sponsor					C
Company employees	C				
Facilities department			C		
Project team				C	
Parents				C	
Children	C				
Childcare staff				C	
Senior management			C		

After a bit more discussion you are pleased that most of the stakeholders are in the desired state. For the few who are not, you develop communication strategies to help move them toward where you would like them to be.

Stakeholder	Unaware	Unsupportive	Neutral	Supportive	Driver
Sponsor					CD
Company employees	C			D	
Facilities department			C	D	
Project team				CD	
Parents				CD	
Children	CD				
Childcare staff				CD	
Senior management			C		D

Additional Information

You can tailor the column titles to represent whatever you feel is appropriate given the needs of your project.

PMBOK® Guide – Sixth Edition References

10.1 Plan Communications Management

10.3 Monitor Communications

13.2 Plan Stakeholder Engagement

13.4 Monitor Stakeholder Engagement

Part 4

Estimating

4.0 ESTIMATING TECHNIQUES

Estimating is a way of life on projects. We are continually estimating durations, resource needs, and costs. This section covers some of the most common estimating methods used in project management, such as analogous, parametric, and three-point estimating.

Forecasts are also estimates, so forecasting techniques such as estimate at completion (EAC) and variance at completion (VAC) are included in this section. The forecasting techniques described in this book use earned value management techniques.

The techniques described in this section include:

* Analogous estimating
* Bottom-up estimating
* Estimate at completion
* Estimate to complete
* Parametric estimating
* To-complete performance index
* Three-point estimating
* Variance at completion

It is a good practice to apply more than one estimating or forecasting technique when possible. If the two estimates are significantly different, you should investigate why they are different to make sure you have identified all the scope, assumptions, and other information necessary to produce a valid estimate.

These estimating and forecast techniques are not the only viable techniques. However, they are the ones most frequently used in project management. Other techniques are used in a generalized business environment, and thus are not described in this book.

4.1 ANALOGOUS ESTIMATING

WHAT IT IS

In its most basic form analogous estimating compares past projects with the current project, determines the areas of similarity and the areas of difference, and then develops an estimate based on that information.

A more robust application determines the drivers for the estimate and analyzes the relationship between past similar projects and the current project. Drivers can include size, complexity, weight, or other aspects of the component for which you are developing an estimate.

Analogous estimates are often used early in the project at a high level. There is an expectation that there will be a range of estimates and that the range will be progressively narrowed as more detailed information is uncovered. However, you can develop an analogous estimate at any level of detail.

The benefits of using analogous estimates are that they are relatively quick to develop, and they're not very costly to develop. However, because they're usually done at a high level, they're not the most accurate method.

HOW TO USE IT

Use the steps below as a guideline. Tailor the steps as necessary to work within your environment.

1. Identify a previous, similar project or projects. Make sure the projects are really similar, not just apparently similar.
2. Determine the aspects of the project that are most likely to influence the cost or duration (drivers).
3. Compare those aspects to the current project and derive a multiplier or divisor to reflect the difference between the projects.
4. Multiply the information from the previous project by the multiplier for each driver (or the project as a whole) to determine the analogous estimate for the current project.
5. Sum the component parts, review for accuracy, and refine as needed to reflect the current project.

Scenario: You are the project manager for a project to implement a childcare facility for your organization's employees.

Duration Estimate Example

You need to develop training materials for the childcare center staff. The training will take two days. You developed a similar training program a few months ago, but that program was three days long. To develop the estimate you follow these steps.

Step 1. Reduce your estimate by 33 percent to account for the shorter training duration.

Step 2. Because this new training program is relatively complex compared with the earlier one (it includes first aid, food prep, and safety training), increase the duration by 10 percent for a complexity factor.

Step 3. The first training class required four hands-on demonstrations, but this one only needs three. You spent 40 hours developing the demonstrations in the last project; therefore, you reduce that amount by 25 percent to end up with 30 hours.

This table shows how you developed the estimate.

Last Class	Effort	This Class	Modifier	Modified Estimate
3 days	200 hours	2 days	–33%	133
Easy	Included	Complex	+10%	13
4 demos	40 hours	3 demos	–25%	30
Total	240 hours			176 hours

Cost Estimate Example

To get a high-level cost estimate for remodeling the space you can look for information from a remodel that your organization recently did and adjust those estimates to fit the current project. If the company doesn't have that history, your contractor will surely be able to pull from his expertise and recent experience. The scenario might sound something like this:

We recently took 5,000 square feet of new space that hadn't been built out at all and put in a childcare facility. That project ended up costing $550,000. But, for this project, we have to demo the existing space, so that will add about 10 percent. However, this project is only 4,000 square feet, so that's 20 percent less space.

Step 1. Add the 10 percent for demo. That brings the cost to $605,000.
Step 2. Multiply $605,000 by 80 percent to get an analogous estimate of $484,000.

Additional Information

Analogous estimating is the most frequently used form of estimating, even if we don't know we are using it. We automatically look to past experiences to help us estimate current projects. This method can be used to estimate durations, costs, and resources.

To use this method effectively, projects must be similar in fact, not just in appearance. Remodeling an office building in New York City and remodeling an office building with the same footprint, in Olympia, Washington, are going to have very different logistics requirements and the cost of resources will be very different.

PMBOK® Guide – Sixth Edition References

6.4 Estimate Activity Durations
7.2 Estimate Costs
9.2 Estimate Activity Resources

4.2 BOTTOM-UP ESTIMATING

WHAT IT IS

Bottom-up estimates are detailed estimates. They include information about all associated costs with an activity or a work package. Bottom-up estimates can include such things as technical requirements, engineering drawings, effort duration estimates, rate estimates, and other direct and indirect costs. These are used to determine the most accurate estimate possible. Bottom-up estimating is the most accurate form of estimating because of the level of detail associated with the estimate. However, you can employ this type of estimating only toward the end of planning, when you have detailed information about the scope. For large projects, bottom-up estimating is a very time-intensive and therefore costly process.

Some cost categories that are included in a bottom-up estimate include:

- Travel
- Certification
- Shipping and handling
- Licenses
- Regulatory requirements
- Legal requirements
- Permits
- Security costs
- Inflation estimates
- Cost of money
- Exchange rates
- Procurement costs
- Contingency and management reserve
- Consultant costs

HOW TO USE IT

Use the steps below as a guideline. Tailor the steps as necessary to work within your environment.

1. Determine the labor effort for each deliverable.
2. Identify the labor rate for each deliverable.
3. Multiply effort hours times the labor rate.

4. Determine the physical resources you need:
    a. Equipment
    b. Material
    c. Supplies
    d. Locations
5. Identify the rates for each physical resource.
6. Calculate the cost estimate for each physical resource by multiplying the amount times the rate.
7. Identify other direct costs.
8. Identify other indirect costs.
9. Determine the contingency reserve.
10. Aggregate all the labor, material, direct, indirect, and reserve cost estimates

---

Scenario: You have been asked to meet the physical growth needs of Top Dog Project Services.

You are working with the Pennsylvania office to lease a building in a new office park to accommodate the expansion of the company's workforce. You are almost done with the Organizing Phase of the project and are just putting the cost baseline together. You have worked with the architecture firm and the general contractor to develop a detailed cost estimate for the Design Phase. The architecture bottom-up costs are presented here.

LABOR	Labor Hours	Labor Rates per Hour	Total Labor
Sr. Architect	175	$225	$39,375
Jr. Architect	600	$165	$99,000
Administrative	250	$55	$13,750
Project Manager	150	$125	$18,750
General Contractor	100	$115	$11,500
Total Labor Costs			$182,375
PHYSICAL RESOURCES	Amount	Rate	Total Resources
Equipment	0		$—
Material	0		$—
Supplies		$6,550	$6,550
Locations	0		$—
Total Resource Costs			$6,550
OTHER	Amount	Rate	Total
Direct Costs			
Travel (mileage)	500	$0.55	$275
Permits	5	various	$11,300
Indirect Costs			
Overhead		2%	$4,010
Contingency		5%	$20,451
BOTTOM-UP ESTIMATE			**$224,961**

Additional Information

The bottom-up estimating method is most often used for costs and resources. It can be used for effort and duration estimates for scheduling; however, to determine the duration of the project the network diagram must be developed and the critical path identified.

PMBOK® Guide – Sixth Edition References

6.4 Estimate Activity Durations
7.2 Estimate Costs
9.2 Estimate Activity Resources

4.3 ESTIMATE AT COMPLETION

WHAT IT IS

Estimate at completion (EAC) is a forecasting technique that estimates the total funding needed for the project, deliverable, control account, or work package. In essence, the EAC is the estimate to complete (ETC) plus any expenditures already incurred. To find out more about ETC, read Section 4.4. This technique builds on the information presented for the ETC.

Here are the abbreviations for the earned value terminology used in this technique:

PV = Planned value
AC = Actual cost
EV = Earned value
BAC = Budget at completion
ETC = Estimate to complete
EAC = Estimate at completion
CPI = Cost performance index (EV/AC)
SPI = Schedule performance index (EV/PV)
CSI = Cost-schedule index (CPI × SPI)

For more information on earned value analysis, see Section 2.5. For more information on performance indexes, see Section 2.8.

The simplest formula to show the EAC is:

$$ETC + AC = EAC$$

Another way of showing it is:

$$BAC - EV + AC = EAC$$

This formula is used if the current cost variance is not expected to continue. It assumes that any existing variance is a one-time event. Most people do not consider this a valid way of calculating the EAC.

You can get a more accurate estimate by dividing the work remaining by a performance index. The performance index that is most often used is the CPI. Thus, the equation is:

$$\frac{BAC - EV}{CPI} + AC = EAC$$

This formula is used if the current cost variance is expected to continue at the same rate. Some people consider this the best-case scenario EAC. This equation has a shortcut:

$$\frac{BAC}{CPI}$$

Another equation that is used to develop an EAC uses both the cost and schedule performance indexes. When considering cost and schedule indexes you assume that both the cost and schedule performance will impact the total cost of the project. Some people consider this the most likely scenario, others the worst case. The equation for this is:

$$\frac{BAC - EV}{CPI \times SPI} + AC = EAC$$

CPI × SPI can be called the cost-schedule index and is abbreviated as CSI.

HOW TO USE IT

If your original estimates were accurate:

1. Calculate the total value of all work completed on your project (your earned value, or EV).
2. Subtract that amount from the total budget (BAC).
3. Add the actual costs incurred to date (AC).

The resulting amount is your estimate at completion for the project.

If your work remaining will continue at the same cost variance as the completed work:

1. Divide your budget at completion (BAC) by your cost performance index (CPI).

If your work remaining will be impacted by your cost and schedule performance:

1. Calculate the total value of all work completed on your project (EV).
2. Subtract that amount from the total budget (BAC) to get the value of your work remaining.
3. Multiply your cost performance index (CPI) times your schedule performance index (SPI) to get your cost-schedule index (CSI).
4. Divide the value of your work remaining by your cost-schedule index (CSI).
5. Add your actual costs incurred to date (AC).

Scenario: Build Twin Pines Medical Plaza, a new state-of-the-art medical resource center.

This example builds off the example in the earned value analysis technique described in Section 2.5. You have completed the groundbreaking, the foundation, the steel frame, and you are 85 percent done with the shell. Here are the numbers:

Work	Planned Value	Percent Complete	Earned Value	Actual Costs
Groundbreaking	4,500,000	100	4,500,000	4,500,000
Foundation	15,000,000	100	15,000,000	15,874,000
Steel Frame	18,000,000	100	18,000,000	18,952,250
Shell	40,500,000	85	34,425,000	36,105,800
Total	**78,000,000**		**71,925,000**	**75,432,050**

The information in the Total Row indicates the cumulative values to date. The total budget for the construction is $126,000,000.

The EAC, assuming the rest of the values were estimated correctly, is:

$$BAC - EV + AC = EAC$$

$$\$126,000,000 - \$71,925,000 + 75,432,050 = \$129,507,050$$

The EAC, assuming the rest of the work will continue at the same level of cost performance as the work done to date, is:

$$\frac{BAC}{CPI} = EAC$$

$$\frac{\$126,000,000}{.95} = \$132,631,579$$

The EAC, assuming the work remaining will be impacted by your cost and schedule performance, is:

$$\frac{BAC - EV}{CPI \times SPI} + AC = EAC$$

$$\frac{126,000,000 - \$71,925,000}{.95 \times .92} + \$75,432,050 = EAC$$

$$\frac{\$54,075,000}{.87} + \$75,432,050 = \$137,587,222$$

You can see there is a significant variance in the EACs depending on which assumptions you think are valid. Therefore, it is important to do some research into what is happening before you select the EAC you will use.

Additional Information

An EAC is often referred to as the "latest revised estimate," or LRE.

There are many ways to calculate an EAC. The most important idea to keep in mind is that you want to use a model that reflects the reality of your project situation. One way you can fine-tune the model is to change the weighting of the CSI. For example, if you think that both cost and schedule will impact your final costs, but you think the CPI will have a more dominant impact than the SPI, you can weight them, for example, [(CPI x .8) + (SPI x .2)]. You can use any weighting you think is reflective of what is happening on your project.

Another way to fine-tune your EAC is to look at the most recent performance. If you are on a three-year project and the overall project CPI is .92, but the CPI over the past six months is .95, then you might want to use .95, as it is a better indicator of future work performance.

PMBOK® Guide – Sixth Edition Reference

7.4 Control Costs

4.4 ESTIMATE TO COMPLETE

WHAT IT IS

Estimate to complete is a forecasting technique that estimates the funding needed to complete all the remaining work. It can be performed for the entire project or for a deliverable, control account, or work package.

Forecasts are usually developed when you are in the midst of a project. At that point, you usually have more accurate information than you did when you developed your initial estimates. Therefore, the most accurate way to develop an estimate to complete (ETC) is to develop an estimate for the remaining work given what you know at the present time. The new estimate should take into account labor rates, production rates, the project environment, existing staffing, risks, and anything else you know about the project.

Developing a new ETC can be quite time consuming on a large project. Therefore, you may want to use some of the mathematical methods of developing an estimate to complete. The mathematical methods are applied when using earned value analysis and use the same abbreviations. As a refresher, here are the abbreviations for terms used in an ETC calculation:

$$AC = Actual\ cost$$

$$EV = Earned\ value$$

$$BAC = Budget\ at\ completion$$

$$CPI = Cost\ performance\ index$$

For more information on earned value analysis, see Section 2.5. For more information on the cost performance index, see Section 2.8.

The first calculation assumes that your original estimates were accurate and that the future work will continue to be based on accurate estimates. In this case, your estimate to complete would be the budget less the value you have earned (accomplished) so far. This is represented as $ETC = BAC - EV$.

The second calculation assumes that any cost variance you have experienced thus far will continue at the same rate. This is represented as $ETC = (BAC - EV)/AC$. In essence, this forecast calculates the remaining work and divides by the current CPI to develop a forecast of the final cost, assuming the future expenditures will occur with the same efficiency as the past expenditures.

HOW TO USE IT

The steps below are to create a mathematical model of your ETC. If you want to conduct a new bottom-up estimate to complete, see the information for bottom-up estimating in Section 4.2.

If your original estimates were accurate:

1. Calculate the total value of all work completed on your project. Note that the total value of work completed is not how much you paid, but rather the total of the budgeted value of the work completed. In other words, the sum of the budgets for the work you have already accomplished.
2. Subtract that amount from the total budget.

The resulting amount is your estimate to complete the remaining work.

If your remaining work will continue at the same cost variance as the completed work:

1. Complete Steps 1 and 2 above.
2. Divide the value of the work you completed (the earned value) by the cost to complete the work (actual cost): EV/AC. This number is your cost performance index (CPI).
3. Divide the result from Steps 1 and 2 above (value of the remaining work) by the CPI:

$$(BAC - EV)/CPI$$

Scenario: Build Twin Pines Medical Plaza, a new state-of-the-art medical resource center.

This example builds off the example in the earned value analysis technique described in Section 2.5. You have completed the groundbreaking, the foundation, the steel frame, and you are 85 percent done with the shell. Here are the numbers:

Work	Planned Value	Percent Complete	Earned Value	Actual Costs
Groundbreaking	4,500,000	100	4,500,000	4,500,000
Foundation	15,000,000	100	15,000,000	15,874,000
Steel Frame	18,000,000	100	18,000,000	18,952,250
Shell	40,500,000	85	34,425,000	36,105,800
Total	**78,000,000**		**71,925,000**	**75,432,050**

The information in the Total row indicates the cumulative values to date. The total budget for the construction is $126,000.000.

The ETC, assuming the rest of the values were estimated correctly, is:

$$BAC - EV = ETC$$

$$\$126,000,000 - \$71,925,000 = \$54,075,000$$

Notice that you have spent $75,504,150. Therefore, you have $50,495,850 left in your budget, and $54,075,000 worth of work left to accomplish.

The ETC, assuming the rest of the work will continue at the same level of cost performance as the work done to date, is:

$$\frac{BAC - EV}{CPI} = ETC$$

$$\frac{\dfrac{\$126,000,000 - \$71,925,000}{\$71,925,000}}{\$75,432,050} = \frac{\$54,075,000}{.95} = \$56,921,053$$

You have spent $75,432,050. Therefore, you have $50,967,950 left in your budget, and $54,075,000 worth of work left to accomplish.

Additional Information

Earned value management is usually only used on very large projects. However, this method of forecasting is relatively simple. As long as you have good estimates for your work and you can compare them to what the work actually cost, you can develop an estimate to complete.

PMBOK® Guide – Sixth Edition Reference

7.4 Control Costs

4.5 PARAMETRIC ESTIMATING

WHAT IT IS

Parametric estimating uses a mathematical model or statistical relationship to determine cost, duration, or resource needs. Not all work can be estimated this way, but for estimates that have quantifiable estimating relationships, it is fast and easy.

Parametric estimating is easy if you have a simple model such as cost per square foot. It can get more complex as you add more variables. If you set up this kind of estimation model on a spreadsheet, it's relatively simple to change the parameters of your assumptions (such as the number of machines) or basis of estimates (cost per machine).

HOW TO USE IT

Use the steps below as a guideline. Tailor the steps as necessary to work within your environment.

1. Identify a quantifiable estimating relationship for your project.
2. Find an estimating database, an expert in the field, or resource that would have cost or duration ratios for the aspect of your project you are estimating.
3. Apply the ratio or equation to your project and adjust for any unique variables.

Scenario: You are the project manager for a project to implement a childcare facility for your organization's employees.

Duration Estimate Example

You need to paint the childcare center. It is 6,000 square feet. You ask the painting contractor how long it will take. He says a painter can cover about 100 square feet per hour. Therefore, you can assume 60 hours of effort. He plans to have three painters on the job. You divide the 60 hours of effort by 3 resources to determine it will take 20 hours, or the equivalent of 2.5 days.

Cost Estimate Example

To get a high-level estimate to build out the childcare center, the general contractor estimates that this type of job costs about $120 per square foot. You have 6,000 square feet. Thus, you estimate the cost to be $720,000.

Additional Information

You can use parametric estimating for the entire project, such as the cost per square foot to build a home, or for a specific deliverable. Many of the estimating databases you can find online are developed using parametric models. The more data that goes into the model, the more accurate the model is likely to be.

PMBOK® Guide – Sixth Edition References

6.4 Estimate Activity Durations

7.2 Estimate Costs

9.2 Estimate Activity Resources

4.6 TO-COMPLETE PERFORMANCE INDEX

WHAT IT IS

The to-complete performance index (TCPI) indicates how efficiently you must perform for the rest of the project to meet your original BAC, or a specific EAC. It divides the remaining work by the remaining funds. It is the ratio of the "work remaining" to the "funds remaining." The formulas are shown below.

$$\frac{BAC - EV}{BAC - AC}$$

$$\frac{BAC - EV}{EAC - AC}$$

Notice that the first equation uses the BAC in the denominator. Use this equation if you want to determine the efficiency rate at which you need to perform in order to meet your budget.

The second equation uses EAC in the denominator. Use this equation if you want to determine the efficiency rate at which you need to perform in order to meet the EAC number used in the denominator.

The TCPI tells you the cost efficiency you must achieve for the rest of the project to meet the target (either BAC or EAC). In other words, for every dollar you spend you have to get X amount of value, where X = TCPI.

HOW TO USE IT

To calculate the TCPI you will need:

- BAC (budget at completion)
- EAC (estimate at completion, if you are using EAC as a target)
- EV (earned value)
- AC (actual cost)

These terms are defined in Section 2.5.

If you want to achieve your original budget estimate:

1. Subtract your earned value (EV) from your BAC to get the value of the work remaining.
2. Subtract your actual cost (AC) from your BAC to determine the funds remaining.
3. Divide the work remaining by the funds remaining.

This gives you the performance efficiency you must achieve to deliver at the original budget estimate.

If you want to achieve an estimate at completion:

1. Subtract your earned value (EV) from your BAC to get the value of the work remaining.
2. Subtract your actual cost (AC) from your EAC to determine the funds remaining.
3. Divide the work remaining by the funds remaining.

This gives you the performance efficiency you must achieve to achieve the estimate at completion.

Scenario: Build Twin Pines Medical Plaza, a new state-of-the-art medical resource center.

This example builds off the example in the earned value analysis technique described in Section 2.5 and the estimate at completion calculations described in Section 4.3.

At this point in the landscaping project you have the following information:

$$BAC = \$126,000,000$$

$$EV = \$71,925,000$$

$$AC = \$75,432,050$$

$$EAC = \$132,631,579^{[1]}$$

Given this information you first calculate the TCPI necessary to meet the original budget of $126,000,000:

$$\frac{\$126,000,000 - \$71,925,000}{\$126,000,000 - \$75,432,050} = 1.07$$

This means to achieve the BAC, for the remainder of the project, you need to get $1.07 worth of value for every dollar you spend. Given that you are partway through the project, and you are thus far getting .95 for every dollar you spend (the CPI is .95), it isn't likely that you will improve by 12 cents on the dollar from now on.

A more realistic target is the EAC. The TCPI using the EAC is:

$$\frac{\$126,000,000 - \$71,925,000}{\$132,631,579 - \$75,432,050} = .95$$

---

[1]This is the EAC derived using the BAC/CPI equation. It assumes the cost variance will continue, but that the schedule performance will not impact the cost.

(continued)

(*continued*)

   This tells you that if you want to achieve the current EAC, you need to keep performing at the same rate. In other words, if the future performance is consistent with the past performance, the result will be that your EAC will remain realistic. If you perform better, you will deliver for less than the EAC; if you perform worse, you will deliver for more than the EAC.

Additional Information

This is a great equation to see whether your expectations are realistic. Many times those accountable for performing the work, whether they are in-house resources or contractors, tend to have an optimistic bias. They somehow think that performance will improve and they will hit their original target. Because future work is different from past work, it's possible. However, history shows it is highly unlikely. For this reason, I sometimes call this the "Let's get real" index.

PMBOK® Guide – Sixth Edition Reference

   7.4 Control Costs

4.7 THREE-POINT ESTIMATING

WHAT IT IS

This is a great technique to use when you need to account for uncertainty in your estimates. When a lot of uncertainty, risk, or unknowns surround an activity or a work package, three-point estimating will give you a range of outcomes and an expected duration or cost estimate. It is called three-point estimating because you start with three estimates. The first estimate represents the best-case scenario, also known as an optimistic estimate. The next is what you consider the most likely scenario, and the last is the worst-case or pessimistic estimate. Notice we use the word "scenario." This is because you are developing estimates based on an assumed situation. Consider these definitions of "optimistic," "most likely," and "pessimistic:"

Optimistic. The optimistic scenario means that you have all your required resources, nothing goes wrong, everything works the first time, and there are no risks or issues that occur. This is represented as an O, for "optimistic."

Most likely. The most likely scenario accounts for the realities of project life, such as someone being called away for an extended period, work interruptions, things not going exactly as planned, and a few issues occur. This is represented as an M, for "most likely."

Pessimistic. The pessimistic estimate assumes unskilled resources or not enough resources, much rework, delays in work getting done, and multiple risks and issues occur. This is represented as a P, for "pessimistic."

The simplest way to develop a three-point estimate is to sum the three estimates and divide by three. This gives you an average. However, this isn't the most accurate way because it assumes an equal probability that the optimistic, most likely, and pessimistic scenarios would occur, and that's not realistic. In reality, the most likely estimate has a greater chance of occurring than either the best-case or worst-case scenarios. Therefore, a more accurate technique is to weight the most likely scenario and take a weighted average. The most frequent method of calculating a weighted average is:

$$\frac{\text{Optimistic} + 4(\text{Most Likely}) + \text{Pessimistic}}{6}$$

Notice that the numerator has six factors you are adding; therefore, the denominator to calculate the average is 6.

While the distribution equation above is the most frequently used, it is certainly not your only option. If you have historical data, or expertise available that shows that the likelihood of the optimistic, pessimistic, and most likely estimates is different, then use those distributions. For example, you may have historical data that shows the last 20 times you have performed a similar activity the optimistic estimate was valid 4 times, the most likely was closer to accurate 8 times, and the pessimistic estimate was closer to accurate 8 times. In that case, you would use an equation that reflected your experience, such as this:

$$\frac{\text{Optimistic} + 2(\text{Most Likely}) + 2(\text{Pessimistic})}{5}$$

This shows that the most likely and pessimistic outcomes are twice as likely to occur as the optimistic outcome. As with all things associated with math, use the numbers to build a model that reflects what you think is true.

HOW TO USE IT

Use the steps below as a guideline. Tailor the steps as necessary to work within your environment.

1. Interview the people who will be doing the work for which you are developing an estimate.
   a. Talk with them about what the cost or duration estimate would be if everything went well. Discuss what "everything going well" would look like.
   b. Then ask them what they expect will happen, given the current environment and their experience with this type of work.
   c. Last, ask them what the cost or duration estimate would be if there were a lot of challenges and get an idea of what those challenges could be.
2. Determine your weighting factors.
   a. The [optimistic + 4(most likely) + pessimistic]/6 is the most common.
   b. If you prefer an equation that uses a different weighting method, apply that along with your rationale and assumptions for your weighting decisions.
3. Apply the calculations to determine the expected outcome (expected cost or expected duration).

Scenario: Develop an eight-hour in-house training video to prepare employees for an industry certification.

You are going to bring in an employee from another office to use as the talent in the video. You are working up the estimates for the plane fare for the employee to come to the videotaping site. You do some research on plane tickets. Based on the time of year and how far in advance you purchase the tickets, you come up with some estimates.

- If she travels during normal business days (no holidays) and you purchase the ticket three weeks in advance, you can get a cost as low as $300 per round trip. This is the best-case scenario.

- More likely, though, you won't book until two weeks prior to the trip, which will cost $450.
- However, in the past, you've had only 48 hours' advance notice, and if this occurs during a holiday, you're looking at a cost of $1,200 per round trip.
- Using the weighted average, you determine your expected cost to be

$$\frac{\$300 + 4(\$450) + \$1200}{6} = \$550$$

Additional Information

You may see this referred to as a beta distribution or PERT estimating. (PERT stands for program evaluation and review technique.) This was a technique developed in the late 1950s to derive a range of estimates and an expected estimate, given the uncertainty in a project.

You may see the estimates annotated for time or cost, as shown below:

- t_o for "time optimistic"
- t_m for "time most likely"
- t_p for "time pessimistic"
- c_o for "cost optimistic"
- c_m for "cost most likely"
- c_p for "cost pessimistic"

PMBOK® Guide – Sixth Edition References

6.4 Estimate Activity Durations
7.2 Estimate Costs

4.8 VARIANCE AT COMPLETION

WHAT IT IS

A variance at completion (VAC) measurement tells you the difference between your budget at completion (BAC) and the latest estimate at completion (EAC). The equation is

$$VAC = BAC - EAC$$

This tells you the amount of funding you will need over and above the original budget estimate.

HOW TO USE IT

1. Calculate an estimate at completion using one of the equations shown in Section 4.3.
2. Subtract the EAC from your budget at completion.

Scenario: Build Twin Pines Medical Plaza, a new state-of-the-art medical resource center.

This example builds off the example in the earned value analysis technique described in Section 2.5 and the estimate at completion calculations described in Section 4.3.

Using the various EACs calculated in Section 4.3, you can see the VACs:

EAC : $129,507,050

VAC : $126,000,000 − $129,507,050 = −$3,507,050

EAC : $129,507,050

VAC : $126,000,000 − $132,631,579 = −$6,631,579

EAC : $129,507,050

VAC : $126,000,000 − $137,587,222 = −$11,587,222

This shows that there will be a variance at completion from −$3,507,050 to −$11,587,222. With this information you can make choices about whether to fund the variance, reduce scope, or see if there are corrective actions you can put in place to minimize the future variances.

Additional Information

When presenting VAC numbers it is a good idea to present the root cause of the variance and a few recommendations on the best action to take.

PMBOK® *Guide* – Sixth Edition Reference

7.4 Control Costs

Part 5

Interpersonal and Team Skills

5.0 INTERPERSONAL AND TEAM SKILLS

Interpersonal and team skills are some of the most important skills you can have when leading a project. Many of the leadership skills we need to exhibit, as project managers, are interpersonal and team skills. There are many other leadership skills than are discussed in this book. In fact, there are entire sections of bookstores with information on interpersonal, team, and leadership skills. The skills discussed in this book include three of the most important skills for project managers:

- Conflict management
- Decision making
- Problem solving

There are many skills that we do not cover as they are general interpersonal skills that are not used any differently in project management than they are in the business world, such as cultural awareness, negotiation, networking, and political awareness. These skills are not conducive to presenting in the step-by-step style that we have adopted in this book.

5.1 CONFLICT MANAGEMENT

WHAT IT IS

Conflict occurs when two or more people have differing interests, needs, goals, beliefs, or values. Conflict is a natural state of affairs in projects. Conflict management is the deliberate attempt to resolve a conflict productively by minimizing the negative impacts and maximizing the positive aspects of the conflict.

Negative impacts from conflict include:

- **Killing team spirit.** If conflict is not managed constructively, team members may become so distant and dysfunctional that they won't want to interact with others on the team.
- **Reduced communication.** When conflict exists, we often shut down and stop talking or engaging.
- **Team member withdrawal.** Many people are conflict-avoidant. In other words, they will take actions to remove themselves from the conflict, including withdrawing, either from the team or from the situation.
- **Reduced trust.** Often in conflict situations team members will lose trust in their teammates. Reduction in trust also reduces the amount of risk a team member is willing to take, either in the project or among project members.
- **Building animosity and factions.** Reduction in trust can lead to negative team behavior such as building factions or cliques. This behavior can negatively impact team cohesiveness and performance.[1]

There are some positive outcomes associated with conflict, which include:

- **Challenging the status quo.** People get to experience different ways of viewing a situation. In many cases a new way of doing something arises out of a conflict.
- **Involving the team in problem solving.** You can treat a conflict like a problem that needs to be resolved by applying the problem-solving steps discussed in Section 5.3.
- **Minimizing group think.** Often, when a team works together over extended periods of time, they start to think alike and they adopt repetitive behaviors and patterns. This adaptive behavior isn't necessarily bad, but it can be beneficial to challenge the status quo and identify different patterns of thinking.

---

[1] Stackpole Snyder, C., *Manage to Lead: Flexing Your Leadership Style*. Project Management Institute. 2012.

- **Deepened team relatedness.** When the team works through conflicts and problems together it increases the team relatedness and improves the group dynamic among team members.

A widely known conflict resolution model is based on work done by Ralph Kilmann and Kenneth Thomas. [2] The model measures behavior in conflict situations based on two dimensions: assertiveness and cooperation. In other words, the extent to which a person attempts to satisfy his or her own concerns (assertiveness) and the extent to which a person attempts to satisfy the needs of others (cooperation) are the primary traits measured. Based on these two dimensions, there are five modes for resolving conflict: competing, accommodating, avoiding, collaborating, and compromising.

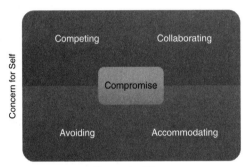

Competing

Collaborating

Compromise

Avoiding

Accommodating

Concern for Self

Concern for Others

Competing is a behavior that is assertive and uncooperative—an individual pursues his own concerns at the other person's expense. This is a power-oriented mode in which you use whatever power seems appropriate to win your own position—your ability to argue, your rank, or economic sanctions. Competing means "standing up for your rights," defending a position which you believe is correct, or simply trying to win.

Accommodating is a behavior that is unassertive and cooperative—the opposite of competing. When accommodating, the individual neglects his own concerns to satisfy the concerns of the other person; there is an element of self-sacrifice in this mode. Accommodating might take the form of selfless generosity or charity, following another person's preference when you would rather not, or yielding to another's point of view.

Avoiding is a behavior that is unassertive and uncooperative—the person neither pursues her own concerns nor those of the other individual. Thus, she does not deal with the conflict. Avoiding might take the form of diplomatically side-stepping an issue, postponing an issue until a better time, or simply withdrawing from a threatening situation.

Collaborating is behavior that is both assertive and cooperative—the opposite of avoiding. Collaborating involves an attempt to work with others to find some solution that fully satisfies their concerns. It means digging into an issue to pinpoint the underlying needs and wants of the individuals involved. Collaborating between persons might take the form of exploring a disagreement to learn from each other's insights or trying to find a creative solution to an interpersonal problem. Collaborating often involves multiple parties. It typically results in a win-win solution.

Compromising is a behavior that is moderate in both assertiveness and cooperativeness. The objective is to find some expedient, mutually acceptable solution that partially satisfies both parties. It falls intermediately between competing and accommodating. Compromising gives up more than competing but less than accommodating. Likewise, it addresses an issue more directly than avoiding, but does not explore it in as much depth as collaborating. In some situations, compromising

---

[2] Thomas, K.W., and Kilmann, R.H., Thomas-Kilmann Conflict Mode Instrument (TKI), Mountain View, CA, 1974.

might mean splitting the difference between the two positions, exchanging concessions, or seeking a quick middle-ground solution.

Each of us uses different conflict management modes based on the situation. We tend to have a dominant mode, one we feel most comfortable with, but we can access the other modes as needed. The following table outlines the five modes of conflict resolution and some situations where each mode can be most effective.

Mode	Situation
Competing	When you need to make an unpopular decision
	When you need to act immediately
	When the stakes are high
	If the relationship is not important
	Be aware that this is usually not an effective long-term method. You should only use this method after other methods have been tried.
Accommodating	To reach an overarching goal
	To demonstrate open-mindedness and flexibility
	To maintain harmony
	When any solution will be adequate
	When you will lose anyway
	To create goodwill
Avoiding	When you can't win
	When the stakes are low
	If you need a cooling-off period
	To preserve neutrality or reputation
	If the problem will go away on its own
Collaborating	When there is time and trust
	When the objective is to learn
	When you have confidence in the other party's ability
	Useful in multicultural situations
	When you need a win-win solution
Compromise	When both parties need to win
	When an equal relationship exists between the parties in conflict
	To avoid a fight

HOW TO USE IT

Use the steps below as a guideline. Tailor them based on the situation and environment.

1. **Clearly articulate the conflict.** You should be able to state each party's position and perhaps the source of the conflict as well. You can't work on resolving the conflict until both parties are clear on the point of disagreement.
2. **Identify rules of engagement.** It is a good idea to set out the rules of engagement, such as not referencing past conflicts, being disrespectful, not interrupting, or taking statements out of context.

3. **Identify what a successful outcome looks like.** Think about what you want out of the conflict resolution. You may not need to both be thrilled with the solution. Perhaps you can agree to settle on a resolution that doesn't meet all your desires, but also doesn't go against any of your key values.
4. **Understand the implications of not resolving the conflict.** There may be situations when time will take care of the situation.
5. **Brainstorm solutions and resolutions.** Similar to problem solving, you should brainstorm and discuss various options to resolve the conflict. There is frequently more than one way to come to a resolution.
6. **Identify any risks, issues, or negative consequences with possible solutions.** Sometimes you may find a solution that both parties can live with, but it introduces downstream risks or problems. Make sure you understand the possible repercussions of each potential resolution.
7. **Select the best option to resolve conflict.** Given what you identified as a successful outcome, and the risks and issues associated with the various solutions, select the best resolution strategy.

Scenario: You have been asked to meet the physical growth needs of Top Dog Project Services.

This example shows how each of the conflict management styles could be applied in various conflicts associated with meeting the demands of the expanding Top Dog locations.

Competing. The Pennsylvania office relocation team has identified four possible locations to build out. The team sent out a survey to get employee feedback on the four locations. They soon realized that no matter which location they chose, there would be some unhappy employees. The project manager presented the results of the survey along with a high-level overview of the pros and cons of each location and a recommended location to the Branch Management Group. After reviewing the information, the Branch Management Group chose a different location than the recommended one.

This competing method demonstrates making an unpopular decision and making that decision when the stakes are high. The Branch Management Group has the position power (authority) to use a competing style of conflict management.

Accommodating. The Seattle office has approached their growth needs by leasing space in a new building and implementing the work-from-home program in their market. The office expansion workgroup wants to start implementing the work-from-home program immediately to alleviate the overcrowding in their current offices. They approach their Branch Management Group with their proposal. The Branch Management Group agrees with the new site for the offices, but informs them that they cannot implement the work-from-home program until the pilot is complete and Headquarters has released a set of company policies supporting the work-from-home program.

The office expansion workgroup is pleased with the approval for the new location, but disappointed that they will not be able to implement the work-from-home program immediately. They use an accommodating response by accepting the decision since they don't have the position power to argue, and they want to keep good relations with the Branch Management Group throughout the project.

Avoiding. The Southern California office is piloting a work-from-home program. One person on the team thinks it is important to ask the employees involved in the program to provide input into the policies and procedures before creating the first draft. Another person on the team thinks they should create a first draft and then send it out to the employees for comment. You are aware of the conflict, but choose not to get involved because you know that either way will work and ultimately the group will end up with a set of policies and procedures that will meet the needs of Top Dog.

Collaborating. The Idaho Falls project manager is concerned about the delayed access to the building they will be remodeling. The City of Idaho Falls is tearing up the street to upgrade the electric, water, and cable for the part of town that will be part of the urban renewal program. Because of all the construction work, they know the downtown traffic will already be a problem and they don't want to make it worse by having construction work on their building take place at the same time.

The project manager for Top Dog is concerned about needing to wait until November to start work because that means they will be working through the winter. Construction work during an Idaho winter is not a good option.

The City Manager's project manager for the urban renewal, the Top Dog project manager, and the General Contractor for Top Dog's renovation and buildout meet to see if they can problem-solve the situation and work out a solution that meets everyone's needs. The Top Dog project manager asks if the City would be open to allowing construction to take place when the City is not working on the streets. The City asks Top Dog to expand on that idea. The Top Dog project manager says that maybe the General Contractor could get a crew that would work on weekends so that they were not causing more construction traffic during the week.

The City says this is a viable alternative, as long as the General Contractor knows that the roads would be inaccessible. The General Contractor could only get materials and equipment in via the back alley because the road would be torn up. The General Contractor states that he has reviewed the site and the access from the alley. It is wide enough to accommodate equipment and trucks, as long as it is blocked off from other traffic.

The City has no problem with blocking the alley during the weekends. The City project manager mentions that they can have their construction crews work four 10-hour days, giving the General Contractor three days per week to have access to the building. The General Contractor says he has some workers who can do three 10-hour workdays. That will allow a 30-hour workweek. It will still take longer than originally hoped for, but they should be able to get enough of the work done to get insulation and windows in before the weather gets too cold. After the City's work on the road is complete, they can resume a normal work schedule.

This example demonstrates cooperative problem solving, and creative collaboration from all involved parties.

Compromise. The Pennsylvania relocation site has been identified. Unfortunately, the new space won't be fully built out until two months after the existing lease expires. The Facilities Manager contacts the existing landlord and informs her that Top Dog will be relocating. He asks her if the company can extend their lease by three months. The landlord is not happy about losing a tenant, but understands Top Dog's growth needs. She says that she can extend the lease, but that the going rate is higher than the rent they are paying now. Rents are about

(continued)

(*continued*)

10 percent higher than the current negotiated rate. She also says that going month-to-month incurs a 15 percent premium on the monthly rate.

The Facilities Manager says that this is a big adjustment. He points out that Top Dog has been an excellent tenant for the past five years. He agrees to the 10 percent higher rate for the three months but asks the landlord to waive the month-to-month fee, and consider this a lease extension rather than a new lease. The landlord agrees, if the Facilities Manager will allow potential lessors access to Top Dog's offices to see if they are interested in leasing them. The Facilities Manager agrees as long as they have 24 hours' notice and there is a Top Dog employee with them while the offices are being shown. The landlord agrees, and they draw up the contract.

This demonstrates give-and-take from both parties. Neither got exactly what they wanted, but they both got some of what they wanted.

Additional Information

To foster the good aspects of conflict and reduce or minimize the negative impacts, try these five steps:

Communicate openly. Often when we are in a conflict situation we stop communicating, either because we don't feel safe communicating or because we develop apathy toward the situation and stop caring about the outcome. Whatever steps you can take to open up dialogue between the parties experiencing conflict will help the situation.

Focus on the issues, not the person. Remember, the conflict is about perceiving situations differently; it is not personal. Keep your behavior respectful while you work to resolve the issue at hand.

Distinguish between interests and position. An interest is the outcome someone wants. A position is the stance they are taking in the conflict. Focus on the outcome each person wants, not the sides or positions they are taking to get the outcome.

Focus on the present, not the past. Stay focused on the current issue rather than thinking about similar situations that have occurred in the past. You may be frustrated because you have engaged in similar situations previously with either the same person or someone else. Bringing up those past situations won't help you resolve the current one; it will only make the situation worse.

Search for alternatives together. By looking for resolutions and alternative solutions together you can help repair any damage the conflict has caused and create a more constructive relationship. In addition, by working together you can generate more creative options for resolution.

PMBOK® Guide – Sixth Edition References

4.1 Develop Project Charter
4.2 Develop Project Management Plan
9.4 Develop Team
9.5 Manage Team
10.2 Manage Communications
13.3 Manage Stakeholder Engagement

5.2 DECISION MAKING

WHAT IT IS

Decision making is selecting a course of action among several alternatives. We make decisions all the time in projects. Many decisions are simple; others are challenging. The challenging ones have better outcomes if you have knowledge of decision-making processes and the factors involved with making good decisions. To make good decisions keep these tips in mind:

Focus on the big picture. Many decisions you make will have an impact on some other aspect of the project, on another project, or on the operations of the organization.

Recognize your own bias and the bias of others. Make sure that neither you, nor the others involved in making the decision, are operating out of either self-interest, or their own biases.

Keep an open mind. Be willing to end up with a decision that you had never considered.

Interpret and analyze data logically by separating facts from opinion. Decisions need to be based on facts, not opinion. Apply critical thinking and good reasoning to ensure you are analyzing information logically and accurately.

HOW TO USE IT

Using a framework can help you make decisions effectively. Use the steps below as a guideline. Tailor the framework as necessary to fit within your environment.

1. **Define the decision accurately, clearly, and precisely.** In some situations, this can be the most challenging aspect of making a decision. Getting your stakeholders to agree on what you are trying to decide and how to articulate it can be difficult.
2. **Define criteria to evaluate data and the course of action.*** Criteria for making a decision can include risk vs. benefit, chance of failure vs. success, ease of implementation, alignment with the project purpose, opportunity for secondary benefits, and so forth.
3. **Challenge assumptions.** Before you make a decision you should consider the assumptions you and others have made around the variables associated with the decision.
4. **Ensure the data you are using to help make the decision is reliable and credible.**
5. **Brainstorm options.** Consider various decisions and their likely outcomes.

6. **Apply criteria.** Once you have numerous options for decisions, you should evaluate how well they meet the criteria you established in Step 2.
7. **Select the best option.** Select an option that fulfills the decision criteria and has an acceptable level of risk and a good chance of success.

*You can use the information in Alternatives Analysis (Section 2.1) to provide a framework for quantifying your decision-making criteria.

Scenario: You have been asked to meet the physical growth needs of Top Dog Project Services.

The Seattle office location is outgrowing its current office space due to the expanding market share and a recent acquisition. Your project is to meet the immediate need of finding better work accommodations for 25 employees and plan for the addition of 50 employees over the next two years. The first phase of your project is assessment and selection. Once you have assessed the options and selected the best one, the remainder of the project will focus on planning and implementation of the selected option.

To make this decision, you put together an office expansion workgroup. This is a small group of people who have a range of knowledge. You want to make sure that the people you will be working with to review the options and make the decisions are open-minded, collaborative, and communicate openly and respectfully. You select a group of people from various departments to help you gather information, define the decision-making criteria, brainstorm, and ultimately make the best decision for your company.

Human Resources. Marge from HR is aware of current trends in the industry from an HR perspective. She has just finished evaluating your organization's work-from-home pilot. She has her finger on the pulse of employee satisfaction at the company, the workforce integration from the acquisition, and the hiring projections for future growth.

Facilities. Bill from Facilities understands the workspace requirements from a space planning perspective. He understands the cost per square foot for employee office space. He would be a lead on any project involving space acquisition, leasing, and buildout.

Operations. VJ brings skills in policies, procedures, employee furnishings, and equipment as well as workflow management.

Marketing and Sales. Janice is aware of the industry trends, the competition, and sales and growth projections. She is tuned into the high employee turnover, as it has negatively impacted her ability to deliver on sales.

Yourself. As a project manager you bring strengths in facilitation, implementation, planning, and big-picture thinking.

Because you have a limited amount of time to recommend a decision to senior management, you set the following expectations with the group:

Meeting 1. Define the decision and establish the criteria to evaluate options.

Meeting 2. Review and challenge assumptions and explore data for reliability.

Meeting 3. Brainstorm options, apply criteria, and develop a recommendation.

Meeting 1

At the first meeting of the office expansion workgroup, you review the growth trends of the company over the past three years, including the two acquisitions. You ask Janice to present information on employee turnover statistics because to meet the growth of 50 additional employees you have to keep in mind the need to backfill vacancies. Janice provides growth projections by product line and over time so you can understand the expected rate of growth and the type of employees you will need to hire.

Bill presents information on the trends for lease and real estate rates in the current location. VJ provides insight on the expected technology advances for employee productivity for the next three years.

After assessing the information, everyone participates in coming up with a succinct and complete decision statement. You end up with the following:

Find the best option to accommodate at least 25 employees within 90 days and allow for the addition of 25 new employees per year for the next two years.

Next you ask each team member to identify the criteria they think are most important for the project and company success. Once everyone has contributed their ideas, you combine them where possible and brainstorm to come up with any additional criteria. You end up with the following list:

1. The solution must address the immediate need to find acceptable space for 25 current employees.
2. The solution should have a positive impact on employee satisfaction.
3. The solution should not increase the current cost per employee from a space perspective by more than 10 percent.
4. The solution should seek to increase integration of employees from the recent acquisition.
5. The solution should not cause a major overhaul of work processes.
6. The solution should not increase bureaucracy.
7. The solution should minimize disruption to the workforce as much as possible.

The group feels good about the work they have done and think that the decision criteria will lead them to a good decision. They agree to think about their own assumptions over the next week. You also ask them to update and validate any quantifiable data associated with lease rates, purchase rates, buildout costs, employee satisfaction data, and market growth projections.

Meeting 2

At the start of the meeting, Bill shares some updated cost data for the lease and purchase prices. The earlier figures were 18 months old. The company is in a fast-growth area and it turns out purchase price rates have risen by 15 percent in the past 18 months. Lease prices have risen by 8 percent. Marge reports that she has verified the employee satisfaction data. Janice adjusts the growth figures for your core product upward by 5 percent. This sales growth could increase the employee growth rates for Years 1 and 2 to 30 new employees per year.

You ask each person for their thoughts and assumptions about the project and the drivers of the project success.

(continued)

(continued)

VJ states that he thinks the high turnover rate is due to paying below-market rate for salaries.

Marge thinks the salary is at market rate. She thinks people are not happy with the working conditions. She stated that the work-from-home pilot she evaluated showed that employees were more productive and happier if they could telecommute two or more days per week.

Janice states that many competitors are able to retain employees by letting them work from home, or offering a 4/40 workweek, even if they are not paying top dollar.

Bill acknowledges that the working environment is not the best. The company has cobbled together offices, furniture, and equipment as they have grown. The existing workspace is clunky and there is a disparity in the work environment depending on which department you work in.

You can see there are differing assumptions that will impact the team's ability to fulfill the criteria of having a positive impact on employee satisfaction. You ask Marge to go through the last 50 exit interviews prior to the next meeting and report on the results.

Meeting 3

Marge presents a summary of the top reasons employees cite for leaving the company.

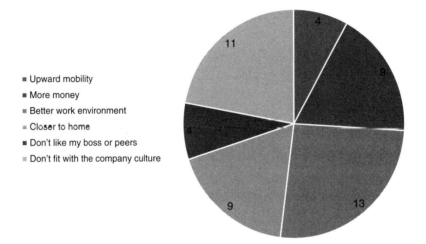

- Upward mobility
- More money
- Better work environment
- Closer to home
- Don't like my boss or peers
- Don't fit with the company culture

She states that the high numbers for not fitting with the company culture are predominately from employees from acquired companies who left shortly after the acquisition was complete.

With this new information you move into a brainstorming session to develop options. After about 15 minutes the group coalesces around three options:

1. Roll out a work-from-home program.
2. Establish a flexible work program.
3. Rent space in a new office park nearby that will accommodate 100 new employees over the next three years.
4. Purchase the existing building, remodel it, and take over another floor that is currently leased to another company.

You apply the criteria to each solution. If the option has a positive impact on the criteria, it is scored a +1. If it is neutral, it is scored 0. A negative impact is scored −1.

Criteria Option	Work from Home	Flexible Work	Rent Space	Purchase and Remodel
The solution must address the immediate need to find acceptable space for 25 current employees.	+1	0	0	−1
The solution should have a positive impact on employee satisfaction.	+1	+1	+1	+1
The solution should not increase the current cost per employee from a space perspective by more than 10%.	+1	+1	+1	−1
The solution should seek to increase integration of employees from the recent acquisition.	−1	−1	+1	+1
The solution should not cause a major overhaul of work processes.	0	0	+1	+1
The solution should not increase bureaucracy.	0	0	+1	+1
The solution should minimize disruption to the workforce as much as possible.	0	0	0	0
Score	2	1	5	2

+1 means a positive contribution.

0 means some or no impact.

−1 means negative contribution.

Based on this outcome you decide that leasing new space is a better option than purchasing the existing building and remodeling. You also decide to recommend an expansion of the pilot work-from-home program to meet the immediate need to find space for 25 employees and to increase employee satisfaction and reduce employee turnover.

Additional Information

As you go through the decision-making process you will need to manage the environment and the team. You should pay particularly close attention to managing the following items:

Challenge assumptions, not people. It is good to challenge assumptions, but make sure it is an assumption you are challenging, and not the person who holds the assumption!

Manage uncertainty. People are uncomfortable around uncertainty. Do your best to introduce certainty where you can.

Manage group dynamics. Making difficult decisions can sometimes bring out the worst in people. Make sure you are managing the dynamics to keep the process constructive.

There are times when it is best to have one person make a decision, and there are other times when a group process is a better option. Keep these guidelines in mind as you determine the best process for the decision(s) at hand.

Individuals generally have the best outcomes in the following situations:

* When time is of the essence
* When group acceptance is not important
* When there is a "best person" who can be identified to make the decision
* When group dynamics are not constructive

Group decision making works well when:

* There is a group of people with diverse knowledge and skills
* There is open communication and acceptance of good ideas
* There are complex problems that require convergent thinking
* The decision needs group buy-in
* The people involved can improve the solution
* The group is effective working together
* There is time to work through the situation

PMBOK® Guide – Sixth Edition Reference

9.5 Manage Team

5.3 NOMINAL GROUP TECHNIQUE

WHAT IT IS

Nominal group technique (NGT) is a method of working with a group to prioritize a list of options. It is often used along with brainstorming to rank the order of ideas, solutions, or options.

In the nominal group technique the various options or solutions are listed. If necessary they are discussed to ensure that everyone is clear on the option or solution. Then each member votes for their preferred option. The votes are tallied and ranked in order of the number of votes.

You can tailor the technique to allow people to vote for their top three options (or any number you prefer). You can also have them rank them using numbers. For example, the first choice is scored as a 5, the second as 3, and the third as 1. Then the numbers are tallied to indicate the ranking of the options.

Nominal group technique is useful because it balances the participation of each person; no single person has a greater say than anyone else. It is a simple method to facilitate group consensus. Because it is objective and transparent it is a fair method of coming to consensus.

HOW TO USE IT

Use the steps below as a guideline. Tailor the steps as necessary to work within your environment.

1. Set up a meeting room with enough chairs for the group, a flip chart, enough 8½ x 11 sheets of paper to record each idea on one sheet of paper, and masking tape. If you want to use sticky notes for voting, have those as well.
2. If you already have a list of options or solutions to prioritize, write each option on a sheet of 8½ x 11 piece of paper and tape it to the wall.
   a. If necessary, engage in decision making (Section 5.2), problem solving (Section 5.3), or brainstorming (Section 1.2) first, to generate a list of options.
3. If needed, summarize each option so everyone is clear.
4. Ask participants to rank their top three options in their preferred order. The highest number goes to the preferred option, the lowest number to the least preferred option.
5. Write each option on the flip chart and tally the results to determine the preferred option.

Scenario: You are the project manager for a project to implement a childcare facility for your organization's employees.

You want the parents to have input into the Childcare Center food service. You posted an electronic brainstorming session over the past week which generated numerous ideas and requirements. Now you are asking the parents to rank the ideas so that you can use them for source selection criteria when evaluating vendors.

The electronic brainstorming session generated 12 requirements overall. You have asked the parents to meet with you to help prioritize the requirements. At the meeting you tape a sheet of paper to the wall for each requirement. You put three ideas per wall on all four walls to give each person a bit of privacy as he or she votes. You have a flip chart that lists each idea in the middle of the room, with the chairs in a semi-circle around the chart. When the parents arrive, you explain each requirement and ask for questions or clarification. Then you describe the voting process.

Each parent is given three sticky notes. One note has the number 3 on it. Each individual is to put that by the requirement that is most important to him or her. The next has a number 2, which is to be placed by the second-most important requirement. The last note has the number 1, which is to be placed by the third-most important requirement.

Once everyone is clear on the process, you tell them to go vote and then return to their seats. Once they vote you ask someone to read you out loud the total for each requirement. You end up with eight requirements that are closely tied, and four that aren't that important. You want to get a clearer division of the top requirements so you ask them to vote on the eight remaining requirements, but this time they can only vote for two requirements. Their top choice they mark with a 3, and their second choice with a 1. You hope this split will create a more obvious preference.

Again, you tally the requirements and you rank the top four as:

a. Healthy food
b. Locally sourced
c. Good references and customer feedback
d. Reasonably priced

You thank the parents for participating and promise to keep them informed on the vendor selection.

Additional Information

You can do multiple rounds of NGT to narrow down a large field of options to a narrower field, and then from the narrow field to the preferred option(s).

PMBOK® Guide – Sixth Edition Reference

5.2 Collect Requirements

5.4 PROBLEM SOLVING

WHAT IT IS

Problem solving is the deliberate act of identifying the problem, identifying and prioritizing alternatives to resolve the problem, and selecting and implementing the solution. The best approach to problem solving depends on the nature of the problem.

A simple problem is one that can be easily defined and the solution can be easily understood and implemented. Simple problems are usually best resolved by one person who is qualified to understand the implications of potential solutions.

A complex problem may be difficult to define, and the solution or decision may also be hard to clarify. There may be multiple stakeholder groups with opposing ideas about what a good solution looks like. The environment may be ambiguous or volatile. All these aspects add complexity to a problem or decision. For complex problems, you get better results when working with a group of people with diverse perspectives.

Another aspect to consider when solving problems is the style of thinking that is most beneficial to solving the problem. Some problems require divergent thinking. This means brainstorming, generating multiple options, and being open to nontraditional ways of approaching the situation. In other situations, a convergent style of thinking is best. In other words, it means narrowing down options and choosing between a limited number of options.

HOW TO USE IT

Using a framework can help you solve problems effectively. These seven steps can be used as a roadmap to guide you through the process:[3]

1. **Define the problem.** The most important step, and frequently the most difficult, is clearly defining and concisely articulating the problem that you are trying to solve. It is important to work with your team, or with others who are trying to solve the problem, to ensure you all have a common understanding and a clear grasp of the problem you are addressing. It can be useful to work with all appropriate stakeholders to come to an agreement on a problem statement. A clearly defined problem statement has two components: (1) what's happening, and (2) the implications of what's happening. A problem should not imply a solution.

[3] Stackpole Snyder, C. *Manage to Lead: Flexing Your Leadership Style.* Project Management Institute. 2012.

2. **Identify the outcomes you want to achieve.** Once you have defined the problem you need to identify what outcomes would resolve the problem. Without clearly understanding acceptable outcomes you won't be able to make the best choices among possible solutions.

3. **Select the criteria you should use to resolve the situation.** If there are certain criteria you need to meet or consider when solving the problem, you should identify those criteria. If there are multiple criteria it may be appropriate to rank them in importance.

4. **Brainstorm solutions.** Once you have done all the prework of defining the problem, identifying the outcomes, and selecting the criteria you can start to brainstorm solutions. Remember, brainstorming should be a free flow of ideas without applying any of the criteria. Brainstorming is a creative process that is used to generate as many options as possible.

5. **Compare solutions to criteria.** Once the brainstorming has stalled, you can start to compare the various options to the criteria you selected. Many of the options will get weeded out in this step. What you are left with is a set of possible solutions.

6. **Identify any risks, issues, or negative consequences associated with each of the possible solutions.** Some of the possible solutions might introduce new risks or issues or they might exacerbate existing risks or issues. After this step is complete you should have at least one viable solution.

7. **Select the best option.** If you have more than one solution, select the one with the highest likelihood of success and the least cost and effort to employ. It is fine to rank the other options for fallback options if needed.

Scenario: You are managing a project to develop a new company intranet site.

The website development project has been steadily falling behind schedule for the past month. At this point, the project manager has called together a group of team members to help solve the problem.

Anthony, the project manager, gives the following information:

Ever since we started the design phase of this project we have been steadily falling behind. We are currently two weeks behind on the critical path and we are coming up on a project review. The technical work on this project is not all that challenging, so that is probably not what is causing the problem.

Anne, the system engineering team lead, says,

Well, we are fully resourced, but most of the designers have been out of school less than five years and they are pretty new to the company. There have been several times when they have taken longer to complete their work than a more experienced designer would take and they have made some mistakes that set the timeline back due to rework.

Bill, the resource manager, states that the original estimates were based on the assumption that all team members would have at least five to ten years' experience and an average skill set and productivity rate.

Anthony says,

I think we can frame the source of the problem as the fact that the people doing the work are different from the people who estimated the work and that the skill sets and expertise are less than estimated. This allows us to state the problem as, "Due to inexperienced team members work is taking longer to complete." Is that correct?

The team agrees with the problem definition and moves into defining the outcomes they want to achieve. Based on a brief discussion they identify the following outcomes:

1. Regain two weeks of the schedule slip.
2. Identify a way to eliminate any future schedule deterioration.

Bill says,

I want to make sure that part of our solution criteria involves upgrading the skills of these team members and does not include any disciplinary action. I think these are good team members, they just need help furthering their skills.

Anne adds that she needs to make sure all the policies and quality checks remain in place. Brian, the finance officer for the project, states that he is comfortable releasing some money from reserves, but that he can't release more than $12,000 to get the schedule back on track.
The team discusses the criteria and ranks it as:

1. All quality policies must be followed.
2. The solution cannot use more than $12,000 in reserve.
3. Team members get help improving their skill set.

Once this is done the four start brainstorming solutions. They continue doing this for about 20 minutes and come up with several possible options. As they compare each option to the solution criteria, a few of them are discarded because they go outside the quality criteria, or are cost prohibitive. What they are left with are the following possible solutions:

* Bring in two outside designers for three weeks.
* Reassign the existing resources to a less critical project and bring in more advanced designers.
* Bring in a team lead with extensive experience to do work and mentor the designers.

These solutions are evaluated for risks and other negative consequences. Based on the current status of the project the group determines that reassigning the existing resources and bringing in a more advanced team would put the schedule further behind and would leave the feeling that the existing resources had failed. It would also not help the existing resources improve their skills.

(continued)

(continued)

After evaluating the two remaining solutions the group decided to bring in a team lead with extensive experience to accomplish some of the more challenging work and to help the existing design team find ways to work smarter and faster. They decided to hold off on bringing in outside designers for the time being but agreed to reevaluate that option if there was not substantial schedule improvement in the next two weeks.

Additional Information

Sometimes people can slip into blaming and complaining when you really need them to stay focused on problem solving. Set up the conversation by framing it as "us versus the problem" rather than pointing fingers.

Be aware of confirmation bias when you are solving problems. Confirmation bias is looking for solutions that fit your own preconceptions. It is important to remain open-minded rather than only listening to information that fits your opinion. A similar bias is called projection. Projection is assuming others share a similar thought belief or value as you do. Projection can be difficult to detect because we often can't see our own beliefs and how they may not be true.

PMBOK® Guide – Sixth Edition References

8.2 Manage Quality
9.6 Control Resources

Part 6

Other Techniques

6.0 OTHER TECHNIQUES

The "other" category of techniques is a catch-all for techniques that don't fit in any of the other categories, but that are important techniques to know about when managing a project. Some of them are foundational techniques that project management is built on, such as precedence diagramming, critical path method, and rolling-wave planning. You will also see some more specialized techniques such as prompt lists, context diagrams, and prototypes.

The techniques described in this section include:

- Context diagram
- Critical path method
- Funding limit reconciliation
- Inspection
- Leads and lags
- Precedence diagramming method
- Prompt lists
- Prototypes
- Resource optimization
- Rolling-wave planning
- Schedule compression

Many of the scheduling techniques described in this section are incorporated into scheduling software. Thus, much of what you will read about happens in the background if you are using software to build your schedule. However, it is always a good idea to know what is happening in the background of your scheduling software so you can troubleshoot your schedule if you need to compress it, or if you think the software isn't behaving correctly.

6.1 CONTEXT DIAGRAM

WHAT IT IS

Context diagrams are often used in system engineering projects and software design projects. They show the system or process, its boundaries, and the people or systems that interact with the process. They provide a high-level overview of the system that can be used to understand the scope and boundaries of a system or process at a glance. Most context diagrams use the following shapes:

A circle represents the process you are modeling.

A rectangle represents systems, processes, or people that interact with the circle.

Arrows show directional flow of inputs or outputs.

HOW TO USE IT

Use the steps below as a guideline. Tailor the steps as necessary to work within your environment.

1. Create a circle in the middle of a chart with the system or process you want to diagram.
2. Identify all the people, systems, and processes that interact with your system. These are called "actors." Actors are roles and not individuals, for example, student, customer, and the like.
3. Draw an arrow that indicates the direction of the information flow from each actor to the process or to other actors.
4. Label the arrows to indicate what flows between the actors and processes.

Scenario: You are managing a project to develop a new company intranet site.

As part of updating your company intranet you are putting in an online ordering application for the cafeteria. This will enable employees to log onto the app, view the menu, place an order, and pay for the order. They can bring their receipt down to the cafeteria and pick up their order in a pick-up area.

You are discussing requirements with the cafeteria management and kitchen staff. They are in favor of the idea because they think it will reduce the wait time and crowding during

(continued)

(continued)

peak hours. The employees like it because they can pre-order their food for a specific time and just pick it up at that time.

The cafeteria manager wants to know if you can use the same application to track the food going out to help manage the inventory. After discussing all the needs, the kitchen staff and management agree that a combination of the sales reports from the system and manual inventory list from the kitchen are needed. Using both these inputs, the kitchen management can place an order with the suppliers who will deliver the food to the kitchen staff.

The group puts together the following context diagram to provide a high-level process chart for the programmers to use when coding the application.

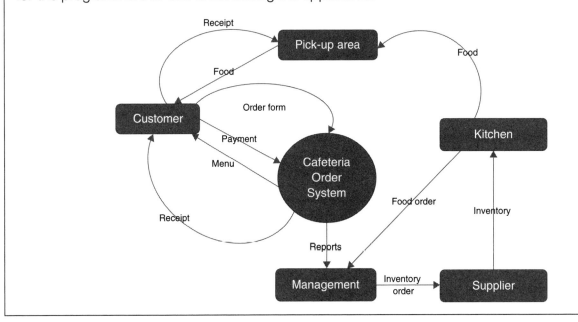

Additional Information

Context diagrams do not show the sequence of steps in a process and they don't require any technical knowledge to understand them.

PMBOK® Guide – Sixth Edition Reference

5.2 Collect Requirements

6.2 CRITICAL PATH METHOD

WHAT IT IS

The critical path method is the most common form of teaching scheduling and is also the method that most software employs. The critical path method determines the date ranges in which activities can occur by calculating the earliest and latest dates activities can start and then the earliest and latest dates activities can finish, based on the network diagram and the activity duration. Most critical path analyses are done prior to loading resource availability into the schedule. After resources are loaded, the duration of the project can change, as duration is based on the resources performing the work. The critical path method has several terms you need to be familiar with.

Early start date (ES). The earliest time an activity can start

Early finish date (EF). The earliest time an activity can finish

Late finish date (LF). The latest time an activity can finish without causing a schedule slip or violating a schedule constraint

Late start date (LS). The latest time an activity can start without causing a schedule slip or violating a schedule constraint

Total float. The amount of time an activity can be delayed from its early start date without causing a schedule slip or violating a schedule constraint

Free float. The amount of time an activity can be delayed without impacting the early start date of the next activity

Forward pass. Working forward through the schedule from the early start date to determine the early start and early finish dates

Backward pass. Working backward through the schedule from the late finish date to determine the late start and late finish dates

Critical path. The longest path through a network, which determines the shortest possible time to complete the project

HOW TO USE IT

Below are the high-level steps to build the schedule, using the critical path method:

1. Draw the network diagram based on the nature of the relationships between the activities.
2. Box the network diagram.

3. Enter durations into the diagram.
4. Conduct a forward pass.
5. Conduct a backward pass.
6. Calculate float.
7. Identify the critical path.

Here is a breakdown on each of the steps.

1. Draw the network diagram.

   For this demonstration we will use the same network diagram used in the Section 6.6 Precedence Diagramming Method.

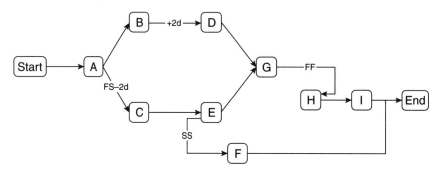

2. Box the network diagram.

   "Boxing the diagram" means setting up the network diagram with a box for each activity and a place to enter the duration, early and late start dates, early and late finish dates, and the float. This prepares the diagram to conduct the forward and backward passes. You can use any structure you want. This table shows one common way of showing the information for each activity.

Early Start	Duration	Early Finish
Activity		
Late Start	Total Float	Late Finish

Using the network above, the boxed network would look like this:

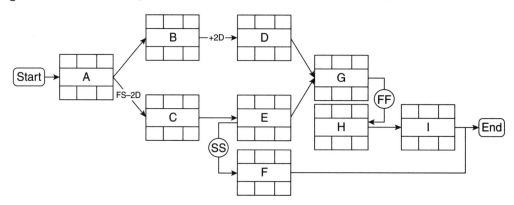

3. Enter durations into the diagram.

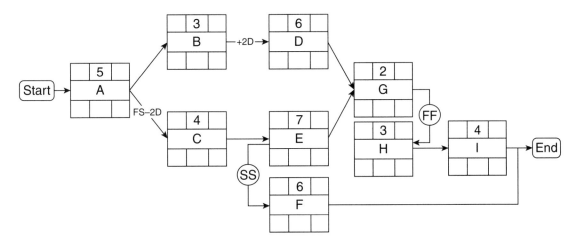

4. Conduct a forward pass.

   The forward pass will tell you the earliest dates activities can start and finish, based on the network logic. Here are the steps to conduct a forward pass:

   1. Set the early start for the first activity or activities to 1.
   2. Add the duration of the activity to the early start, then subtract 1. That is the early finish date. In this example, Activity A starts at Day 1, you add the duration of 5 to that, subtract 1, and the early finish is Day 5. You subtract 1 from the early start plus the duration because you started on Day 1, not Day 0.
   3. Add 1 to the early finish to determine the early start of the immediate successor activities. Note that scheduling software assumes that an activity starts at 8 AM and is complete at 5 PM on its due date. Therefore, the next activity starts the next morning at 8 AM.
      a. For this example, the early start of Activity B is 6. However, the early start of Activity C is 4 because there is a two-day lead. The early start of C would have been 6, but you subtract two days from 6 because of the two-day lead, and that gives you an early start of 4.
      b. To continue along with C, you add the duration of 4 to the early start of 4, subtract 1, and get an early finish of 7.
      c. Going to Activity B, you add the early start of 6 and the duration of 3, subtract 1, and get an early finish of 8.
      d. There is a two-day lag between B and D, so Activity D starts two days after B finishes. This means that D will start on Day 11.
      e. Continue in this fashion until you reach the end of the network. The figure below shows a network diagram with a forward pass completed.

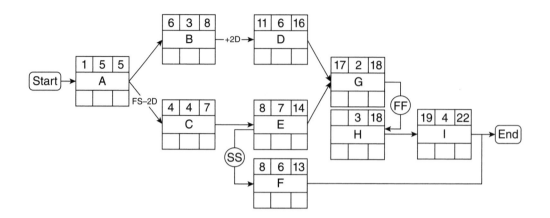

Note these areas of interest in this diagram:

- Activity F has the same early start as Activity E. That is because there is a start-to-start relationship. Activity F can start after Activity E starts.
- Activity G and Activity H have a finish-to-finish relationship. This means that Activity H can finish after Activity G finishes.
- Activity D and Activity E both have to finish before Activity G can start. Activity D has an early finish of 16, and Activity E has an early finish of 14. When there is a path convergence in a forward pass, the highest number always carries forward.

  After you finish calculating your forward pass, you will see the earliest finish date. This is the soonest the project can finish, given the information available. For our example, the early finish date is 22 days.

4. Conduct a backward pass. A backward pass tells you the latest time activities can start and finish without violating a schedule constraint (such as a mandatory milestone) or without causing the project to be late. For a backward pass, follow these steps:
   a. Take the early finish date of the last activity in the network and enter that number as the late finish date as well. Therefore, your early finish and late finish for the last activity will be the same. For this example, it's 22 days.
   b. Subtract the duration and add 1 to establish the late start for the last activity in the project. Because you start the project at Day 1, instead of Day 0, you need to add the 1 day back in to get accurate dates. For this example, 22 minus 4, plus one gives you a late start date of 19.
   c. Using the late start date, subtract 1 to derive the late finish date for any immediately preceding activities. This demonstration has two final activities: Activity I and Activity F. Use the highest number as the late finish date. Activity I finishes on Day 22, and Activity F finishes on Day 13. Therefore, the late finish for the project is Day 22.

d. For activity H the late start date minus 1 carries backward to become the late finish for Activity H, which is 18 in this example.

The figure below shows how the demonstration network looks after a backward pass.

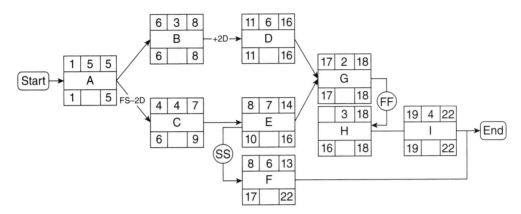

Note these areas:

• Activity H does not have an early start, but it does have a late start. That's because in this scenario, I'm not so much worried about when Activity H can start; rather, I'm more concerned about when it finishes. In a real-life scenario, you would have more information about the nature of the activity in the activity attributes. That would give you insight into whether Activity H would need to start on Day 16, or whether it could start any time.

• The late start of Activity D is 11, but the late finish of its predecessor (Activity B) is not 10, but 8, because there is a two-day lag. You treat that lag as if it were a mini-activity called "don't do any work here." Therefore, you calculate the late start of B as 10 and you subtract 2 from that to get a late finish of 8 for Activity B.

• The relationship between Activity E and Activity F is start-to-start. It would seem that the late start of Activity F would drive the late start of Activity E. However, in this case, the late start of Activity G drives the late finish of Activity E, which in turn determines the late start. This demonstrates an important rule in the backward pass: When you have two possible options for a late start or finish, the lowest one is the one you select.

5. Calculate float.

After you calculate the early and late start and finish dates, you can determine the float. Float is the difference between the late finish and early finish dates.

For finish-to-start relationships, the difference between early start and late start dates will be the same as the difference between the early finish and late finish dates.

However, when you have a start-to-start or a finish-to-finish relationship, you have to determine the difference between the early and late start and the early and late finish, and take the lowest number as the amount of float.

The figure below shows the network diagram with the float entered.

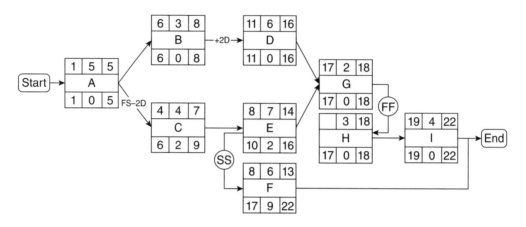

Note these areas:

• Only Activities C, E, and F have total float. Activity F has the most, with nine days of float: It can finish up to nine days later than its early finish date and not cause the project to be late. As I mention earlier, to calculate the nine days of float, you subtract the early finish from the late finish, or the early start from the late start. Activities C and E both have two days of float. However, that float is shared between them. Any float that Activity C uses isn't available for Activity E. If Activity C starts on Day 5 instead of Day 4, then Activity E will start on Day 9 and have only one day of float.

• There is something special about Activities E and F: They have free float. They are the last activities on a path. Therefore, if they start late or finish late, within their float, no other activities are affected. Even though Activity C has total float, if it starts late, then both E and F start late. But if Activity F starts late, it doesn't cause any disruption in any other activity, as long as it finishes by Day 22.

• If Activity E starts late, then F starts late; but if E finishes late, by less than two days, then no other activities are impacted. Therefore, you can see that those activities with free float are the most flexible in the network. You can delay them and reallocate resources to activities that have little or no float to ensure on-time performance, assuming that resources have the right skill sets.

6. Identify the critical path

Finally, identify the critical path, which is the path with the least amount of float. For this example, the critical path is A-B-D-G-H-I. Usually, the critical path is presented as the path with zero float. However, say that you're given a mandated end date, and you calculate your critical path, and you have negative float. That tells you that you cannot meet the end date, given the current information, meaning you have to find ways to compress the schedule or negotiate a new end date.

Conversely, if you have positive float, you have a little time to spare. In both circumstances, the path with the least amount of float is your critical path.

Scenario: You are putting in a new backyard.

This example builds off the network diagram from Section 6.5 Leads and Lags and Section 6.6 Precedence Diagramming Method.

The first step is to box the diagram as shown below.

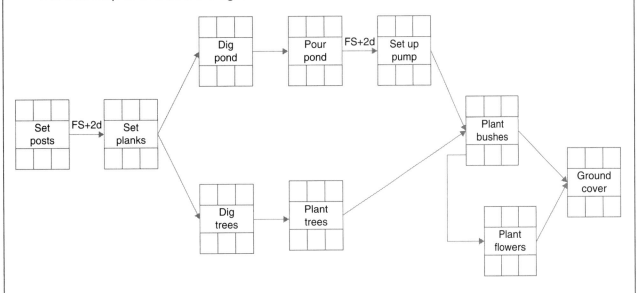

After talking with the landscape architect and the contractor who will be working on the yard, you determine the following durations for the activities:

Set deck posts	1 day
Set deck planks	3 days
Dig pond	2 days
Pour pond foundation	1 day
Set up pond pump	1 day
Dig tree holes	2 days
Plant trees	2 days
Plant bushes	1 day
Plant flowers	1 day
Spread ground cover	1 day

(continued)

(continued)

You add these durations into the network diagram.

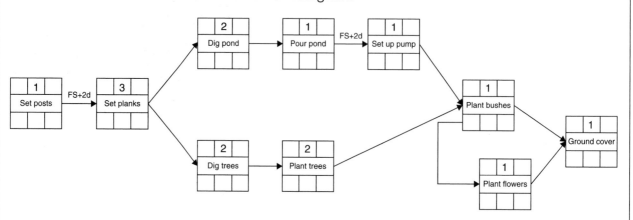

Next you conduct a forward pass to determine the early start and finish dates. This is how the network looks with a forward pass:

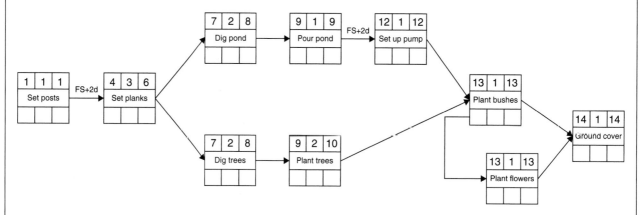

Notice that the two-day lag impacts the early start of the next activity.
Then conduct a backward pass to determine the late start and finish dates.

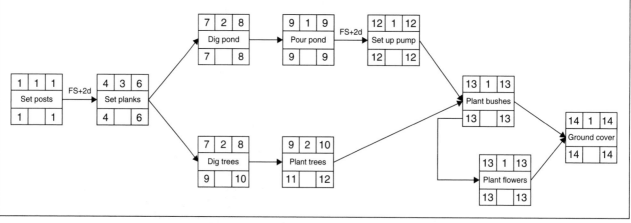

The next step is to calculate the float.

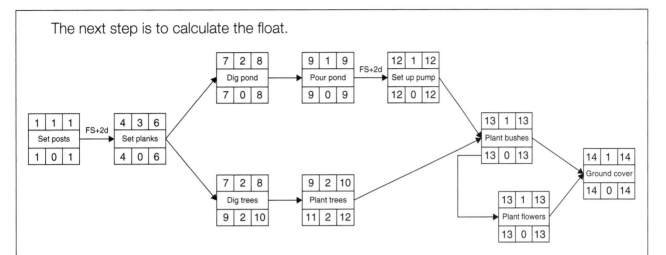

There are only two activities with float, digging the tree holes and planting the trees. They each have two days of float. This float is shared between them. If it takes one day longer to dig holes for the trees, then planting the trees only has one day of float remaining. We say this as "float belongs to the path, not the activities."

This means that the critical path consists of these activities:

- Set deck posts
- Set deck planks
- Dig pond
- Pour pond foundation
- Set up pond pump
- Plant bushes
- Plant flowers
- Spread ground cover

Additional Information

The "+1 day/–1 day method" is a bit clunky to emulate when you're doing a manual calculation. To make it easier, use Day 0 as an early start for the first activity; that way, you don't have to worry about adding or subtracting days based on the start or end of the business day. But be careful, the early start date and late start date will be different than if you use the +1/–1 method. The early and late finish dates will be the same and the float will be the same. PMI's Project Management Professional Certification (PMP) uses the +1 day/–1 day method.

PMBOK® Guide – Sixth Edition References

6.5 Develop Schedule
6.6 Control Schedule

6.3 FUNDING LIMIT RECONCILIATION

WHAT IT IS

Funding limit reconciliation is used to align the flow of work with the availability of funds. Some organizations fund projects based on an annual budget. All planned work must be accomplished in a budget year, and you can't do work that isn't in that year's budget. Therefore, before you finalize your schedule and your budget you need to make sure you have scheduled the work so it occurs when the funds are available.

HOW TO USE IT

Use the steps below as a guideline. Tailor the steps as necessary to work within the funding constraints of your organization.

1. Develop your initial schedule of work.
2. Develop your initial budget for the work.
3. Determine any funding limitations, such as funds that aren't available until a certain time, or funds that must be expended before a cut-off date.
4. Identify any inconsistencies on when work is scheduled and when funds are available.
5. Reschedule the work to coincide with the funding availability.

Scenario: You have been asked to meet the physical growth needs of Top Dog Project Services.

You work for the Idaho Falls location of Top Dog Project Services. Your office has outgrown its existing location. The state is promoting an urban renewal program for Idaho Falls. The City is upgrading the infrastructure to support high-tech companies including high-speed cabling and mini-cell towers. To incentivize companies to move their businesses here, they have offered tax breaks and reduced permitting fees. They are also providing a $25,000 grant for companies that incorporate green building principles in their architecture and design. To take advantage of this program all planning and design documents must be submitted by September 5. In addition, work on the building cannot start before November 1 because the City will have the streets torn up to put in the new infrastructure.

Your sponsor is the Chief Operating Officer. He thinks this is a great way to make a positive impact on the community and save money. He has asked you to look at your schedule and come back with options to meet the funding constraints.

This is an excerpt from your schedule:

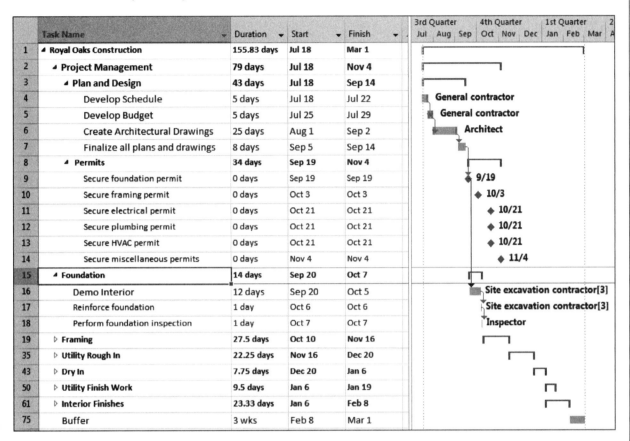

	Task Name	Duration	Start	Finish
1	◢ Royal Oaks Construction	155.83 days	Jul 18	Mar 1
2	◢ Project Management	79 days	Jul 18	Nov 4
3	◢ Plan and Design	43 days	Jul 18	Sep 14
4	Develop Schedule	5 days	Jul 18	Jul 22
5	Develop Budget	5 days	Jul 25	Jul 29
6	Create Architectural Drawings	25 days	Aug 1	Sep 2
7	Finalize all plans and drawings	8 days	Sep 5	Sep 14
8	◢ Permits	34 days	Sep 19	Nov 4
9	Secure foundation permit	0 days	Sep 19	Sep 19
10	Secure framing permit	0 days	Oct 3	Oct 3
11	Secure electrical permit	0 days	Oct 21	Oct 21
12	Secure plumbing permit	0 days	Oct 21	Oct 21
13	Secure HVAC permit	0 days	Oct 21	Oct 21
14	Secure miscellaneous permits	0 days	Nov 4	Nov 4
15	◢ Foundation	14 days	Sep 20	Oct 7
16	Demo Interior	12 days	Sep 20	Oct 5
17	Reinforce foundation	1 day	Oct 6	Oct 6
18	Perform foundation inspection	1 day	Oct 7	Oct 7
19	▷ Framing	27.5 days	Oct 10	Nov 16
35	▷ Utility Rough In	22.25 days	Nov 16	Dec 20
43	▷ Dry In	7.75 days	Dec 20	Jan 6
50	▷ Utility Finish Work	9.5 days	Jan 6	Jan 19
61	▷ Interior Finishes	23.33 days	Jan 6	Feb 8
75	Buffer	3 wks	Feb 8	Mar 1

The first problem you notice is that you aren't scheduled to have the plans finalized until September 14th. The next problem is that you are scheduled to start the building demo on September 20. Thus, you will have to accelerate the design phase and delay the buildout phase to align with the grant funding and the availability of the new infrastructure.

In discussing this with your sponsor you decide to fast-track some of the drawings to overlap with the detailed planning for the schedule and budget. You will also incentivize early delivery. If the architecture firm can deliver final plans and drawings you will give them a $5,000 bonus.

Additional Information

Funding limit reconciliation is often a factor to consider when working on federal projects, as there are strict guidelines about funding work within the appropriate budget year. Some federal projects also have constraints about the types of funds you can use, such as capital project funds or operational funds.

PMBOK® Guide – Sixth Edition Reference

7.3 Determine Budget

6.4 INSPECTION

WHAT IT IS

Inspection, when used in the context of project management, is used both to verify that the deliverables or results are correct, and as a method to validate the deliverables or results as acceptable to the customer. In some situations verification and validation occur at the same time; in other situations, the verification that the product or result is technically compliant is part of a quality control process before releasing the product for customer acceptance.

There are four primary types of inspection:

Test. For quantifiable requirements and things you can measure. This is good for size, weight, and speed.

Examine. For requirements you can verify by a visual inspection. Used when you can examine the deliverable to verify that it meets requirements.

Analysis. This method is used if you can't see what is happening, but you can infer from the result that things work correctly.

Demonstration. For deliverables that need to perform one or more steps, you can demonstrate the steps.

It is a good idea to conduct both verification and validation activities throughout the project. Verification efforts are focused on assessing that the product or solution has been designed and developed according to the specifications. Validation efforts focus on assessing whether the right product or solution has been developed. You can perform inspections as deliverables are completed, at phase gates, or at milestones. This alleviates the risk that you will get the finished product and the customer will say, "This is not what I was wanting at all!"

HOW TO USE IT

Because inspection is product specific it is impossible to accurately describe how to inspect an individual product. This description provides a high-level overview of the process rather than a detailed description.

1. Gather the product acceptance criteria, requirements documentation, statement of work, specifications, or any other standard or documentation that contains a detailed description of the deliverable or result.

2. Determine the best way to test or prove that the deliverable meets the criteria. Common methods include testing, examining, analysis, and demonstration, depending on the nature of the deliverable.
3. Conduct the inspection, review, or audit.
4. Record the results. If the deliverable or result is not compliant, you may need to perform a root cause analysis (Section 2.11), correct a defect, or submit a change request.

Scenario: You are the project manager for a project to implement a childcare facility for your organization's employees.

For the childcare center, you are following the practice of validating the results with the key stakeholders (the parents) throughout the process. At the start of the project you meet with the parents to gather their requirements. You use the following points along the way to gain validation:

* You develop a prototype (see Section 6.8 for information on prototypes) of the layout to show parents what the childcare center space will look like.
* When the blueprints are complete, you show the parents how the blueprints meet the requirements and match the prototype.
* When the rough construction work is done, you do a walkthrough with the parents to verify that construction is consistent with the blueprints.
* When the childcare center is complete you do a final walkthrough to gain acceptance that the final buildout is acceptable.

There are also specific deliverables in the childcare center that you verify for correctness along the way:

* You can test how the security system works by making sure the cameras cover the area you need and that the alarm company is notified when the perimeter is breached.
* You can examine the furniture and equipment to make sure that it's child-size and safe.
* You can analyze the enrollment software to see that it performs as promised.
* You can demonstrate that the security system cameras and motion detectors show up in the correct locations on the computer screens and that they correctly represent the physical layout of the childcare center.

Additional Information

Inspections are sometimes called audits, walkthroughs, or reviews. Many industries and professions have their own terminology to convey the same concept.

PMBOK® Guide – Sixth Edition References

5.5 Validate Scope
8.3 Control Quality
12.3 Control Procurements

6.5 LEADS AND LAGS

WHAT IT IS

Leads and lags are used to refine your network diagram and thereby your project schedule. They modify the relationship between activities. A lead accelerates the timing between two activities and a lag delays the timing between two activities. You can use leads and lags with any type of relationship (finish-to-start, finish-to-finish, and so forth).

Here are some examples:

Finish-to-start (FS) with a lead: *Two weeks before I am done gathering the requirements, I will start designing the center.*

Note that the lead is demonstrated with a minus (–) sign.

Finish-to-finish (FF) with a lag: *I will finish editing the User Instructions five days after I finish writing them.*

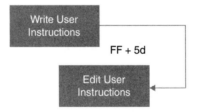

Note that the lag is shown as a plus (+) sign. This indicates that the "finish editing User Instructions" is dependent on finishing writing them. If the end date for writing slips, the end date for editing slips.

Start-to-start (SS) with a lag: The predecessor activity must start before the successor activity can start. As an example, *I can start writing test questions before I can start writing test answers.*

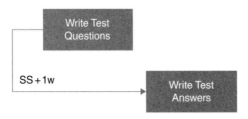

This indicates that you will start the test answers a week after you write the questions. You don't have to finish the test questions before you start on the test answers. If you delay the start of the test questions, you will delay the start of the test answers.

The time indicators are often abbreviated as:

M   Month
W   Week
D   Day

HOW TO USE IT

These steps describe how to modify a precedence diagram using leads and lags manually. You can do the same type of actions using scheduling software. Scheduling software uses the abbreviations described in this section and in Section 6.6 Precedence Diagramming Method.

1. Analyze your network diagram, looking for places where you can overlap activities by modifying the type of relationship and adding a lead or a lag.
   a. Are there places you can accelerate a finish-to-start relationship with a lead?
   b. Are there places you can change a finish-to-start relationship to a start-to-start with a lag?
2. Once you have modified the dependencies and included leads and lags, review the network again to see if you have introduced risks by modifying the flow of the activities. Too much overlap in activities increases risk.

This is an example of the same network diagram shown in the precedence diagramming method, only there is an example of a lead and a lag. The lead is between activities A and C and the lag is between B and D.

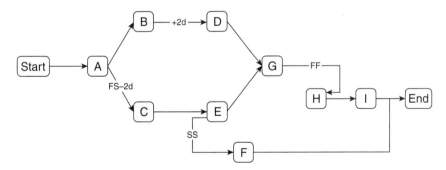

Scenario: You are putting in a new backyard.

You looked at your initial network diagram for the backyard and determined that you had not allowed any time for the deck posts to set or for the pond foundation to cure. You add two days of lag to both of those activities.

(continued)

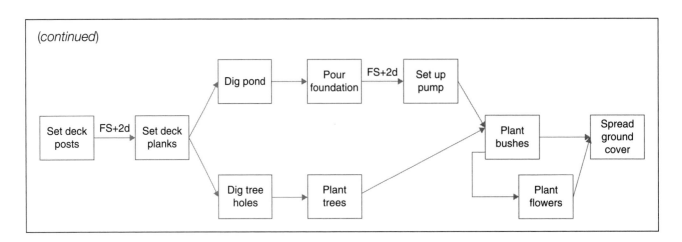

Additional Information

When you modify the relationships to do activities in parallel that are normally done in sequence, you are employing a schedule compression technique called fast-tracking. The schedule compression technique is described in greater detail in Section 6.11.

PMBOK® Guide – Sixth Edition References

6.3 Sequence Activities
6.5 Develop Schedule
6.6 Control Schedule

6.6 PRECEDENCE DIAGRAMMING METHOD

WHAT IT IS

The precedence diagramming method (PDM) is the preferred method of establishing logic relationships between activities to create a network diagram. The network diagram is one of the key components in developing a schedule. In PDM each activity is represented as a box, connected to other boxes with an arrow to indicate the sequence in which activities occur. Boxes are often referred to as nodes in scheduling terminology.

There are four types of relationships used in PDM:

Finish-to-start (FS): This is the most common type of relationship. The predecessor activity must finish before the successor can begin. For example, *I have to finish gathering requirements before I start the design.*

Finish-to-finish (FF): The predecessor activity must finish before the successor activity can finish. A sample scenario is: *I have to finish writing the User Instructions before I can finish editing them.*

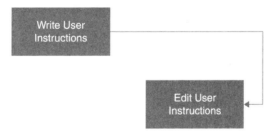

Start-to-start (SS): The predecessor activity must start before the successor activity can start. As an example: *I need to start writing test questions before I can start writing test answers.*

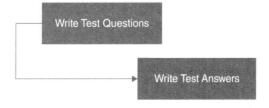

Start-to-finish (SF): This is the least common relationship. You will not see this very often, but I include it here to be complete. The successor activity must start before the predecessor can finish. Consider this scenario: A new accounting system must be operational before the legacy system can be decommissioned.

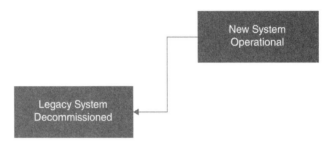

The relationships are often abbreviated as:

FS   Finish-to-Start
FF   Finish-to-Finish
SS   Start-to-Start
SF   Start-to-Finish

HOW TO USE IT

These steps describe how to develop a precedence diagram manually. You can do the same type of actions using scheduling software. Scheduling software uses the abbreviations shown above.

1. Start with a list of all your activities.
2. Create a start node.
3. Identify the first activity or activities.
4. Place the first activity after the start node.
5. Determine the next activity and place it after the first activity.
6. Continue until all the activities are placed in the precedence diagram (aka network diagram).
7. Review to determine if this is optimal, or if some of the relationships should be different (for example, if some of the finish-to-start relationships should be start-to-start).

This is an example of a network diagram developed using the precedence diagramming method. Note there is a start-to-start relationship between activities E and F, and a finish-to-finish relationship between G and H.

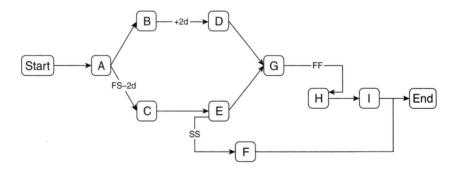

Scenario: You are putting in a new backyard.

To turn your new backyard from a set of plans into reality, you identify the following activities:

1. Set deck posts
2. Set deck planks
3. Dig pond
4. Pour pond foundation
5. Set up pond pump
6. Dig tree holes
7. Plant trees
8. Plant bushes
9. Plant flowers
10. Spread ground cover

   After looking at the work to be done and the resource constraints (you only have two workers) you decide to do the deck first, then the pond and the trees at the same time. After those are done you will work on the bushes and the flowers at the same time, and then finish with the ground cover. You develop a network diagram that looks like this:

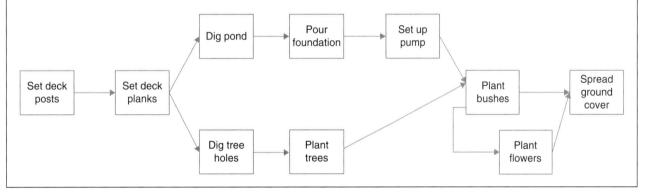

Additional Information

A viable schedule usually has about 80 to 90 percent of the activities in a finish-to-start relationship.

   In the beginning, when you are creating a high-level overview of the project schedule you may want to start out with sticky notes so you can experiment with the flow of work and deliverables. Once you are confident you have the high-level flow correct, you can enter the information in software and create your network diagram electronically.

   Many times, the relationships in the preceding list are required based on the nature of the work, but some relationships are more discretionary and some relationships are based on external constraints. The nature of the dependency impacts the flexibility you have in your schedule. Here are the different types of dependencies.

   Mandatory. These are required based on the nature of the work involved or are required by contract, regulations, or legal restrictions. You can't test something that is not yet developed. For example, the rough electrical work (wiring throughout the space) that has to be done before the finish electrical (hooking up lights, switches, and outlets).

Discretionary. These relationships are determined by a best practice or resource availability. A best practice example is: "I recommend gathering all the requirements before designing the solution, but you can do some design without having the requirements complete and finalized." The relationship is discretionary and based on the choice.

External. An external dependency is based on something or someone outside the project. Requiring a permit, waiting on a vendor delivery, or needing a deliverable from another project are all examples of external dependencies.

Internal. An internal dependency is based on a dependency within the project. These can be based on resource availability or a preference in how to sequence the work. If the same resource is scheduled to complete two tasks at the same time, the decision on which to do first is an internal dependency.

You can see that mandatory and external dependencies are constraints on how you schedule a project. Discretionary and internal dependencies have the greatest flexibility in your schedule. However, you can still find ways to adjust mandatory dependencies by applying leads and lags as described in Section 6.5.

PMBOK® Guide – Sixth Edition Reference

6.3 Sequence Activities

6.7 PROMPT LISTS

WHAT IT IS

Prompt lists identify high-level categories of risks. They are used to help ensure that all areas are considered when identifying project event risks and overall project risks. A prompt list is not exhaustive; rather it is used to help focus attention on different categories of risk so that risk identification is robust and thorough.

Different industries and professions should have different prompt lists. For example, an aerospace firm would not have the same types of risks as an environmental cleanup organization. The following generic list can be further decomposed into subcategories that are appropriate for your project to make it more robust.

- Technical innovation
- Legal, compliance and regulatory
- Quality
- Requirements
- Contracting strategy
- Contractors
- Political
- Environmental
- Stakeholders
- Team resources
- Physical resources
- Stability
- Technical difficulty
- Project complexity
- Schedule
- Budget
- Estimating
- Assumptions and constraints
- External events
- Ambiguity

- Uncertainty
- Volatility
- Social

Some of these categories are more applicable to overall risk, such as volatility, ambiguity, and uncertainty. Other categories are more applicable to event risks, such as team resources, stakeholders, schedule, and budget.

HOW TO USE IT

Prompt lists are often used in a risk management meeting to assist in brainstorming when identifying risks.

1. If your organization has a prompt list, start with that. If not, you can use the list above and tailor it to the needs of your project and organization.
2. Ask meeting attendees to review the list prior to the meeting so they are familiar with the categories.
3. Ask attendees if there are any additional categories they think should be added to the list.
4. Review each item on the prompt list and ask, "What are the risk events or conditions that could occur in the area of _____?" (Fill in the blank with each prompt.)
   a. Some of the categories are more applicable to overall project risk rather than individual risks. For those, substitute, "What are the areas of overall project risk associated with _____?"
5. At the end of the meeting, update the prompt list with any revisions that have occurred as a result of the meeting.

Scenario: You have been asked to meet the physical growth needs of Top Dog Project Services.

You work for the Seattle location of Top Dog Project Services. Your location has two projects to accommodate growth: expanding the work-from-home program and leasing new space to build out. You develop two different prompt lists for your meeting because, though some of the risk categories are the same, there are many that are unique to each aspect of the project.

The first half of the meeting delves into the risks associated with the work-from-home program. You sent out a prompt list that included the following categories:

- Legal, compliance and regulatory
- Requirements
- Experience
- Assumptions and constraints
- Uncertainty

You point out that "uncertainty" is an overall risk. You start the meeting by asking if anyone has something to add. The business analyst on the team points out that the organization currently has no processes around working from home and she thinks that is an entire category of risks. You agree and add "Process" to the list.

The team spends about an hour brainstorming risks and ends up with a viable start to a Risk Register.

After a break you call the meeting back to order and focus on the risk categories associated with the new space buildout. Your prompt list includes the following categories:

- Legal, compliance and regulatory
- Requirements
- Contractors
- Assumptions and constraints
- Estimating
- Schedule
- Budget

The team agrees with the list. They spend 90 minutes brainstorming risks and develop a robust set of risks for the Risk Register.

After the meeting you update the prompt list to make sure the "process" category is included for future reference.

Additional Information

If your organization does not have a formal prompt list, the lowest level of the work breakdown structure or the lowest level of the risk breakdown structure is a good place to start.

PMBOK® Guide – Sixth Edition Reference

11.2 Identify Risks

6.8 PROTOTYPES

WHAT IT IS

A prototype is a model of an end product or process. It is usually created early in the project to give stakeholders an opportunity to visualize the end product. A prototype is generally created based on the known requirements, which allows stakeholders to "get a feel" for the end product. This aids in getting feedback on requirements or deliverables before expending a lot of time, effort, and expense. Examples of prototypes include:

Simulations. Simulations are sometimes called virtual prototypes. They are often developed prior to a physical prototype model using CAD (computer-aided design) or CAE (computer-aided engineering) software. You can model component parts, or entire products, using simulations. Simulations are a relatively inexpensive way to help stakeholders uncover their expectations and requirements.

Three-dimensional small-scale models. These are often used for construction projects so stakeholders and customers can see the design in context and get an idea of the relative scale and size of each component.

Storyboards. Storyboards are drawings that show examples of, or the progression of, a product or result. They are used in software development, computer gaming, film, and in some commercial real estate projects.

HOW TO USE IT

Because prototyping is product specific it is impossible to accurately describe how to prototype an individual product. This description provides a high-level overview of the process rather than a detailed description.

1. Once you have an initial set of requirements, determine the best method to create a prototype: simulation, model, or storyboard.
2. Create your prototype.
3. Show it to stakeholders and record their feedback.
   a. Get clarification on more detailed expectations and requirements.
   b. Identify any changes to requirements.
   c. Refine what stakeholders like and dislike about the prototype.
4. Repeat this process until you have a solid understanding of stakeholder needs and requirements.

Scenario: You are the project manager for a project to implement a childcare facility for your organization's employees.

The childcare center has 3,000 square feet to build out. Your stakeholders have identified that they want the following requirements to be part of the center:

1. An enclosed drop-off and pick-up point
2. An art area
3. A music area
4. A reading area
5. An inside play area
6. An eating area
7. A food prep area
8. A quiet time and nap area
9. Playground equipment
10. Open space outside for running or group games

Before you start developing blueprints you want to get more feedback from the parents on their preferences. You decide to have an open house with several prototypes for the parents to look at and discuss. You use the following prototypes to help elicit more detailed requirements:

Storyboards. You have several drawings of the outdoor space with different configurations of playground equipment and the open play area.

Simulations. You develop several CAD designs of the inside space.

Based on the feedback from the open house you develop a model of the design that had the most favorable feedback. You incorporate a few changes to align with some additional suggestions. You bring the model to a meeting with the parents to gain some final feedback prior to developing the blueprints.

Additional Information

Prototypes can be used as an inspection technique (see Section 6.4) to assist in validation throughout the project life cycle.

PMBOK® Guide – Sixth Edition Reference

5.2 Collect Requirements

6.9 RESOURCE OPTIMIZATION

WHAT IT IS

Resource optimization is done after the initial critical path is identified. (Section 6.2 describes the critical path method.) At that point, you load your resources into the scheduling tool. Often, you will find that one or more resources are over-allocated. For example, you may have scheduled a resource to work full time on more than one activity at the same time. Another example is if you have a resource for a limited amount of time (ten hours per week), but you have scheduled the resource for more time than that. Sometimes you can resolve these resource allocation issues by using float to optimize the schedule without extending the due date. However, the reduction of float also increases the number of critical or near-critical activities. Other times, you will find that the schedule due date is pushed out, and you have a new critical path based on resource availability.

Resource optimization adjusts the start and finish dates to make the best use of your resources. There are two ways to optimize your resources; you can level them, or you can smooth them.

Resource leveling adjusts activity start and finish dates so that resources are no longer over-allocated. Leveling frequently pushes out the due date. For example, if Sarah is scheduled to work on Activity A and Activity B during the week of March 1 through March 5, and both activities require five days of full-time commitment, resource leveling would extend the start date of Activity B to begin on March 8 and end on March 12.

Another option is to find an available resource who has similar skills as Sarah. If that person is available March 1 through March 5, you can level the work across two resources.

Another purpose of resource leveling is to reduce peaks and valleys in resource utilization. Instead of working 40 hours one week and 10 the next, resource leveling seeks to have a resource work 25 hours each week.

Resource smoothing is similar to resource leveling, but it adjusts activities within their float amounts, therefore it does not change the critical path. Smoothing resources is achieved by delaying some work or activities. This reduces the flexibility of the schedule when it comes to dealing with delays but it is a cost-effective way of managing and utilizing resources. In some cases, some resources may remain over-allocated.

HOW TO USE IT

Use the steps below as a guideline. Tailor the steps as necessary to work within your environment.

1. Evaluate your schedule for any instances of resource over-allocation. Most software allows you to view your resources and has an indicator if a resource is over-allocated.
2. Determine if the over-allocated resources are on the critical path, or if they are working on activities that have float.
   a. If the activities are on the critical path you will need to either extend the schedule or bring in additional resources with the same capabilities.
   b. If the activities have float, and the resource has the availability, push the work out until the resource is not over-allocated, or until you reach the point where that work will become the new critical path.
3. Look for uneven work patterns to see if there is a way to level out the work commitment.

Scenario: Develop an eight-hour in-house training video to prepare employees for an industry certification.

You are developing the content for the video in-house. You have several resources working on the content. One resource, Lynne, is accountable for two modules. The current schedule has her working on both modules at once.

		M	T	W	TH	F
Module 1	Lynne	8	8	8	8	8
Module 2	Lynne		8	8	8	8

You level the schedule so that she is only working on one module at a time, which extends the schedule by three days.

		M	T	W	TH	F	M	T	W
Module 1	Lynne	8	8	8	8	8			
Module 2	Lynne						8	8	8

You have another resource, Ethan, whose schedule is a bit erratic.

		Week 1					Week 2				
		M	T	W	TH	F	M	T	W	TH	F
Module 3	Ethan	8	8	2	2	8					
Module 4	Ethan						8	8	2		

You level out his hours so that he has a more consistent workday.

		Week 1					Week 2				
		M	T	W	TH	F	M	T	W	TH	F
Module 3	Ethan	6	6	6	6	6					
Module 4	Ethan						6	6	4		

Additional Information

Working overtime for short periods of time (no more than two weeks) can be an acceptable option as long as there are no regulations or policies that prohibit the extra hours. If you do ask people to work overtime, be aware that there may be a cost involved. Once you start asking people to work more than 50 hours a week or more than six days in a row their productivity rate starts to drop. Thus you will need to make sure you keep the overtime to a minimum to gain the most benefit.

If you are seeking to level work across multiple resources, make sure they have the skill set to do the work. Just because someone is available, doesn't mean he or she has the capabilities necessary to accomplish the work.

If you are bringing additional resources, make sure you account for the time it will take to get them trained on the content and the processes needed to do the work. Sometimes the extra time spent bringing a new person up to speed and coordinating work negates the time you make up by bringing him or her onto the project.

PMBOK® Guide – Sixth Edition References

6.5 Develop Schedule
6.6 Control Schedule

6.10 ROLLING-WAVE PLANNING

WHAT IT IS

Rolling-wave planning is a type of progressive elaboration where you plan the work in the near future (up to about 90 days out) in detail, and leave the work beyond that at a higher level. This doesn't mean you don't know or understand the scope, and it doesn't mean you don't have scheduled deliverables or cost estimates. What it means is that you elaborate the deliverables into activities and assign resources closer to the time you will need them.

Developing a detailed schedule and budget can be very time consuming. Along the course of a project many things change. At the start of a project, there are a lot of aspects of it that are not known. Spending time trying to get a detailed level of knowledge about work that is 6, 9, or 12 months in the future is often not the most productive use of your time. Therefore, keep a schedule that is updated to reflect detail for the near future and as time passes, fill it in with additional detail as it becomes known.

HOW TO USE IT

Use the steps below as a guideline. Tailor the steps as necessary to work for your project.

1. Planning begins with your work breakdown structure. You should have an idea about all the high-level deliverables at Level 2 or 3, depending on the size of your project.
2. Decompose the deliverables that are in the near future to a lower level of detail.
3. Transfer this information into a schedule, if it is not already there.
4. Decompose the work into activities, assign resources, and establish start and finish dates for the activities.
5. Repeat this process throughout the project and update the information with more detail as it becomes available.

Scenario: You are the project manager for a project to implement a childcare facility for your organization's employees.

At the start of the childcare center project you have a WBS that looks like this:

CHILDCARE CENTER

1. Requirements
   1.1 Curriculum
   1.2 Enrichment

(continued)

(continued)

   1.3 Play
   1.4 Operations
2. Plans
   2.1 Interior
   2.2 Outdoors
   2.3 Operations
3. Construction
   3.1 Interior
   3.2 Outdoors
4. Operational Readiness
   4.1 Staffing
   4.2 Food services
   4.3 Curriculum
   4.4 Administration
5. Project Management
   5.1 Startup
   5.2 Organization
   5.3 Management
   5.4 Closeout

About 90 days before you are scheduled to begin work on the outdoor construction (3.2 on the WBS) you start to fill in more detail. That part of the WBS evolves to look like this:

3.2 Outdoors
   3.2.1 Design
   3.2.2 Landscape
      3.2.2.1 Trees
      3.2.2.2 Plants
   3.2.3 Equipment
      3.2.3.1 Slide
      3.2.3.2 Climbing apparatus
      3.2.3.3 Swings
      3.2.3.4 Sandbox
      3.2.3.5 Water area
   3.2.4 Surfaces
      3.2.4.1 Pads
      3.2.4.2 Cement

About 60 days prior to the start of the work, the equipment (3.2.3 on the WBS) should look like this:

3.2.3 Equipment
   3.2.3.1 Slide
      3.2.3.1.1 Determine needs
      3.2.3.1.2 Conduct market research

3.2.3.1.3 Order slide

3.2.3.1.4 Slide delivery

3.2.3.1.5 Install slide

This shows how the scope has not expanded; it is just filled in with more detail as the start of the work gets closer.

Additional Information

You can set your waves at whatever increment works best for your project. For a six-month project you may use increments of two months. For a project that is three years, you may plan four to six months in the future. When you are getting ready to start a new phase of the project life cycle, it is a good idea to update and elaborate your existing schedule and budget.

PMBOK® Guide – Sixth Edition Reference

6.2 Define Activities

6.11 SCHEDULE COMPRESSION

WHAT IT IS

Compressing a schedule involves looking for ways to shorten the overall project duration without reducing the project scope. The two main types of schedule compression are crashing and fast-tracking.

Crashing

Crashing looks for cost/schedule trade-offs. In other words, you look for ways to shorten the schedule by applying more resources, having resources work longer hours, or by spending more money. The intent is to get the most schedule compression for the least amount of money. Some common ways of accomplishing this include

- **Bringing in more resources.** Sometimes more people working, or using additional equipment, can speed up progress. Be careful. Sometimes adding resources actually extends the duration because coordination, communication, and conflict actually take more time than they save!
- **Working overtime.** Many times, you can work longer hours or work weekends. However, this is useful only for a couple of weeks. After that, people burn out and are less productive, so this is a short-term fix. You also need to take into consideration any union or labor regulations.
- **Paying to expedite deliverables.** This can include overnight shipping and paying bonuses to contractors for early delivery.

Crashing usually increases cost, so weigh the benefit of earlier delivery against cost. Bottom line: You're looking for the most time for the least amount of cost.

Fast-Tracking

Fast-tracking shortens the schedule by overlapping activities that are normally done in parallel. You can overlap all or part of a task. One way of doing this is changing the network logic by using leads and lags (see Section 6.5). For example, you can change a finish-to-start relationship to a finish-to-start with a lead. This causes the successor activity to start before the predecessor is complete. You can also change it to a start-to-start or finish-to-finish with a lag.

Another version of fast-tracking is breaking up long tasks into smaller chunks and overlapping the smaller chunks. For example, if you are documenting a manual for a new product, rather than have the documentation as one long activity and the editing as another long activity, you can break them into thirds and overlap them. While one resource is writing the second part of the documentation, another resource can edit the first part.

Make sure you understand the real relationship between activities when you fast-track, or else you will end up with a schedule that doesn't make sense and can't be executed.

Demonstration

Here is a demonstration of a three-week schedule with five activities. The following schedule shows eight hours of work per day.

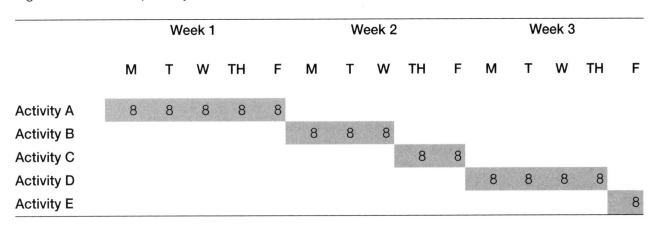

The following version of the schedule shows fast-tracking.

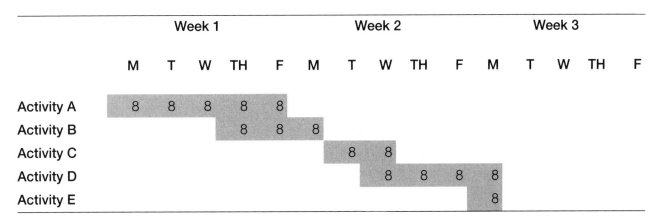

- The relationship between A and B changes from a finish-to-start relationship to a finish-to-start with a lead.
- The relationship between C and D changes from a finish-to-start relationship to a finish-to-start with a lead.
- The relationship between D and E changes from a finish-to-start relationship to a finish-to-finish relationship.

The following version of the schedule shows crashing.

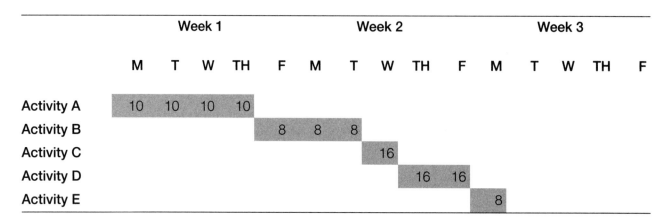

- Activity A used overtime of two hours per day to accomplish the work in four days instead of five days.
- Activity C used two people instead of one.
- Activity D used two people instead of one.

Depending on the nature of the work and the environment, you can crash and fast-track to end up with a viable schedule.

HOW TO USE IT

When you are looking to shorten the schedule, start by looking at the critical path. If you are using scheduling software you can sort the activities by the amount of float, from least to greatest. That will help focus your attention on the activities that you can compress to either recover from a schedule slip or to create some buffer (reserve) in your schedule. If you see any activities that have negative float, that is an indicator that you are behind on the critical path.

Crashing

1. Identify the activities with zero or negative float.
   a. If you have some activities where you can add available resources at the same rate, you may end up cost neutral. For example, if you have an activity that takes 40 hours with one person, but only 20 hours with two people, you will still only pay for 40 hours of work, but get the work done faster.
   b. Look for activities where you are waiting on a delivery. If you can expedite the delivery for an acceptable cost, you can reduce your schedule duration.
   c. Evaluate overtime costs. If you can work overtime for a limited amount of time, the premium you pay for overtime may be worth the time you make up.
2. Start by crashing the activity on the critical path where you can make up the most time for the least cost.
3. If you need to crash further, find the next activity where you can make up the most time for the least cost.
4. Continue until you have met your schedule target or until you can no longer crash any activities for an acceptable cost.

Fast-Tracking

1. Look for activities that use separate resources and can occur at the same time. You can overlap these and not cause a resource constraint.
2. Determine if you can do the activities completely in parallel, or if you should set them up as:
   a. Finish-to-start with a lead
   b. Finish-to-finish with a lag
   c. Start-to-start with a lag
   d. Any other version of dependency and leads/lags
3. Continue until you have met your schedule target or until you cannot fast-track any activities for an acceptable risk.
4. Review your schedule and determine if the risk incurred by overlapping is acceptable. Determine if you need to take any additional actions to reduce risk.

Scenario: You have been asked to meet the physical growth needs of Top Dog Project Services.

You work for the Idaho Falls location of Top Dog Project Services. You are remodeling an old downtown building as part of an urban renewal program. The City is upgrading the infrastructure and a lot of the streets will be torn up while new utility infrastructure is put in place. You were scheduled to start demolishing the interior on September 20. However, with the new infrastructure you cannot gain access until November 1. You want to make up some of the delay to get the building ready for move-in as soon as possible. Here are your options:

1. You can crash some activities by bringing in an extra carpenter, electrician, plumber, HVAC person, or painter and reduce the duration. Bringing in more resources might even reduce the time without increasing cost because the extra resources will shorten the duration.
2. You can also ask those resources to work more hours (depending on any labor union restrictions). Overtime is usually billed at time and a half, so that option may cost more.
3. Your original schedule was set up with finish-to-start dependencies between blueprint activities, carpentry, electrical, plumbing, and HVAC. You can decide to conduct the electrical, plumbing, and HVAC activities in parallel instead of sequentially.
4. You can change a finish-to-start relationship in plumbing to a finish-to-start with a lead. After the bathrooms have all the pipes, the fixtures can be installed. The finish work doesn't have to wait until the kitchen or the outside plumbing is roughed in.
5. You can install the windows and doors after the rough carpentry is done by modifying the relationship to a finish-to-finish relationship with a lag.

Additional Information

You should only crash work on the critical path. It is of no use to spend money to shorten a noncritical path because you won't accelerate the delivery. You can't crash everything. For example, you can't pay to have two inspectors to do one occupancy permit inspection. And you can't pay them to expedite the process.

Fast-tracking can increase your risk and impact quality on the project and might necessitate rework.

PMBOK® *Guide* – Sixth Edition References

6.5 Develop Schedule
6.6 Control Schedule

Appendix: Case Study Scenarios

Every technique described in this book has an example of how it can be applied. It is not possible to provide one case study that allows each technique to be demonstrated. Thus, I created eight different case studies to provide a variety of projects to show how each technique is applied. Since readers come from all different backgrounds, I have tried to stay away from overly detailed and technical examples and explanations. So even though there are construction and IT examples, they are not so specific that you have to be in that industry to understand what is happening.

The case studies are listed here in the order in which they appear in the book.

Information Technology Help Desk

Every quarter your organization conducts an employee satisfaction survey for the internal support services. The support services are comprised of human resources, information technology, legal, maintenance, and security.

For the past several quarters, satisfaction with the IT department has been declining. The Chief Operating Officer wants you to identify what is causing the decline and what can be done about it.

> **Scenario: Your project is to help improve customer satisfaction with the phone support from the IT Help Desk.**

Intranet Website Development

Your organization has grown from 100 people to over 500 people in the past five years. To accommodate the growth you have had to put new systems, work flows, and policies in place. New hires used to shadow existing workers to get trained but now no one seems to know all the latest information, and it seems to be changing weekly to accommodate the expansion.

The organization has put together monthly department meetings to maintain a friendly working environment. These meetings have become the de facto way of finding out about new policies and work flows. What used to be a fun meeting seems to be all about bureaucracy and rules.

The manager of IT suggested putting together an intranet to store all policies, procedures, processes, and work flows. She said it can also be used for newsletters, awards, department updates, announcements, and other information. This idea had a positive reception and you have been assigned as the project manager.

Scenario: You are managing a project to develop a new company intranet site.

Childcare Center

Your organization's employees are primarily young families with children. Forty percent of the workforce is younger than 35 and most have toddlers and pre-school age children. Three of the organization's core values are:

1. Work-life balance
2. Valuing employees
3. Creating a supportive work environment

The Executive Management Group has decided to reinvest some of the profits back into the company by building and managing an onsite childcare center for employees' children. Management believes that in addition to aligning with the corporate values, this will increase employee satisfaction and employee retention, and reduce work-loss time due to childcare issues.

The project includes identifying requirements, managing the construction (which will be outsourced), ensuring adequate indoor and outdoor play options, and engaging with stakeholders throughout the project.

Scenario: You are the project manager for a project to implement a childcare facility for your organization's employees.

Company Growth Needs

Top Dog Project Services is a company with several offices throughout the country. The company growth rate is holding steady at 5 to 10 percent per year. With this rate of growth, several of the offices are outgrowing their current office space. Your project is to work with Headquarters and each office to identify options to meet the office space needs for the growing workforce for the next five years.

You have met with the Division Directors and identified the following offices as having the most urgent needs:

* Idaho Falls. Idaho Falls is looking at moving their office to an area of downtown that is undergoing an urban renewal program. There are some perks being offered to companies that participate early, though there are risks involved as well. The City has a requirement to keep the brick building facades to maintain a classic feel to the downtown area.
* Southern California. The Southern California office is pushing to develop a work-from-home pilot program that can be expanded to other locations in the future.
* Seattle. The Seattle location plans on moving to a different office building. They would like to adopt the work-from-home program as a competitive necessity in their market.
* Pennsylvania. The Pennsylvania office has a lease that is expiring in eight months. They plan on leasing and building out an office in a new business park.

Scenario: You have been asked to meet the physical growth needs of Top Dog Project Services.

In-House Training Video

Your organization wants to support employees in getting certified in their field. The Human Resources department has surveyed employees about their intention to pursue certification. Based on the responses, your organization expects 150 employees to take the certification over the next four years.

Your organization has 50 employees who have already taken and passed the exam. Several of them have volunteered to help develop the material needed to prepare people to pass the exam. This idea has met with favorable feedback. However, management is concerned that these employees will spend too much time training and that it will impact their work.

Your Marketing Director mentions that she has worked with a video company that specializes in in-house training products. They can use their own talent, or your company employees to make the video. Senior Management has decided to use internal employees who have already passed the exam as subject matter experts (SMEs) to develop the material and be in the video. You have been assigned to research the costs of using the video company versus filming it using in-house resources. Once that decision is made, you will manage the project.

Scenario: Develop an in-house training video to prepare employees for an industry certification.

PMO Upgrade

You work for a project management consulting organization. Your organization has decided to update and upgrade your current PMO information system infrastructure with all new software, cloud computing, collaboration sites, and real-time reporting software.

The expectation is that you will market this platform for a monthly subscription fee to use all the features. You currently have 12,500 subscribers for various project management services. The project will take one year to complete, so the income will not be realized until next year.

Scenario: Your organization is updating its current PMO information system infrastructure with all new software, cloud computing, collaboration sites, and real-time reporting software.

Twin Pines Medical Plaza

The City of Twin Pines has grown over the past decade from 12,000 to 60,000 people. The nearest hospital and medical facilities are 40 miles away. In a rare act of collaboration, the city, county, and state have put together a bond and tax-funding stream to build a new state-of-the-art medical resource center called Twin Pines Medical Plaza (TPMP). Twin Pines Medical Plaza will have:

- Hospital
- Medical Center with various family care physicians and specialists
- Urgent Care Center
- Assisted Living Community
- Community Center with programs for the community such as:
  - Weight management
  - Smoking cessation
  - Addiction recovery
  - Nutrition
  - Fitness programs

The following is a work breakdown schedule based on these high-level phases.

Twin Pines Medical Plaza

1. Alternatives Development
   1.1 Alternatives analysis
   2.2 Conceptual engineering
2. Reviews
   2.1 Environmental impact
   2.2 Traffic analysis
   2.3 Noise assessment
3. Design
   3.1 Preliminary design
   3.2 Permitting
   3.3 60% design
   3.4 90% design
4. Construction
   4.1 Ground breaking
   4.2 Foundation
   4.3 Steel frame
   4.4 Shell
   4.5 Interior frame
   4.6 Trades
       4.6.1 Electric
       4.6.2 Plumbing
       4.6.3 HVAC
       4.6.4 Gas
       4.6.5 Alarm
       4.6.6 Telephone
       4.6.7 Cable
   4.7 Finishes

5. Equipment
- 5.1 ER/Trauma
- 5.2 Surgical
- 5.3 Cardiac
- 5.4 Neuro
- 5.5 CCU
- 5.6 ICU

6. Staffing
- 6.1 Administrative
- 6.2 Trauma
- 6.3 Surgery
- 6.4 Specialist
- 6.5 Technicians
- 6.6 Nursing
- 6.7 Support
- 6.8 IT
- 6.9 Business office

Scenario: Build Twin Pines Medical Plaza, a new state-of-the-art medical resource center.

New Backyard

After years of talking about it, you have decided to put in a new backyard. You are going to transform what is now a fenced lawn into a wonderful, peaceful environment where you can barbeque, garden, and sit outside and relax. You have the plans drawn up for a new deck, a pond, some fruit trees, bushes, flowers, and ground cover.

Scenario: You are putting in a new backyard.

Index